Enhancing Interoperability and Automation of Construction Waste Quantification

Enhancing Interoperability and Automation of Construction Waste Quantification outlines the construction waste quantification (CWQ) modelling that supports data-driven decision-making in the built environment. It presents how the functionalities of Building Information Modelling (BIM) and Semantic Web Technology are integrated to enhance the interoperability and automation of the CWQ process.

The research in this book shapes the development of a semantic framework that supports the built environment in quantifying construction waste (CW) and informing optimal material choices from early design stages to minimise the quantity and diversity of waste generation. The book also demonstrates the application of the proposed framework using an ontology (PROduct CIRcularity Ontology) and a BIM-integrated digital tool (Building Waste Tool [BWT]). The PRODCIRO and BWT inform how data standardisation, consistency, and granularity could streamline and automate the CWQ process. The book also presents the outputs of a test-case building used to validate the adaptability and accuracy of the framework.

This book is a valuable resource for BIM and sustainability practitioners. It provides a comprehensive discussion on the significance of CW, its impacts on sustainability, advancements in CWQ, and data and information gaps within the existing CWQ practices. The solution proposed in the book will help the built environment to shift from reactive to proactive and preventive construction waste management.

Subarna Sivashanmugam is Research Associate at the School of Computing, Engineering and Digital Technologies, Teesside University, UK.

Sergio Rodriguez Trejo is Associate Professor of Research at the School of Computing, Engineering and Digital Technologies, Teesside University, UK.

Farzad Pour Rahimian is Professor of Digital Engineering at the School of Computing, Engineering and Digital Technologies, Teesside University, UK.

Digital Frontiers in Buildings and Infrastructure: Navigating the New Era of Digital Engineering
Series Editors: Farzad Pour Rahimian and Mohamed Kassem

This series provides comprehensive coverage of both technical innovations and strategic applications of digital technologies in the built environment and civil engineering sectors.

Structural Design and Optimization of Lifting Self-Forming GFRP Elastic Gridshells Based on Machine Learning
Soheila Kookalani, Hamidreza Alavi and Farzad Pour Rahimian

Enhancing Interoperability and Automation of Construction Waste Quantification
Integration of Building Information Modelling and Semantic Web Technology
Subarna Sivashanmugam, Sergio Rodriguez Trejo and Farzad Pour Rahimian

Enhancing Interoperability and Automation of Construction Waste Quantification

Integration of Building Information Modelling and Semantic Web Technology

Subarna Sivashanmugam, Sergio Rodriguez Trejo and Farzad Pour Rahimian

Routledge
Taylor & Francis Group
LONDON AND NEW YORK

First published 2025
by Routledge
4 Park Square, Milton Park, Abingdon, Oxon OX14 4RN

and by Routledge
605 Third Avenue, New York, NY 10158

Routledge is an imprint of the Taylor & Francis Group, an informa business

British Library Cataloguing-in-Publication Data
A catalogue record for this book is available from the British Library

ISBN: 978-1-032-99854-1 (hbk)
ISBN: 978-1-041-00609-1 (pbk)
ISBN: 978-1-003-61076-2 (ebk)

DOI: 10.1201/9781003610762

Typeset in Times New Roman
by Apex CoVantage, LLC

Contents

1 Introduction

1.1 Built Environment and Construction Waste

Due to urbanisation and increasing population, raw material consumption and solid waste generation increase at an alarming rate. In particular, housing, construction, and infrastructure consume 40–50% of the global raw materials and generate half of the global solid waste annually [1]. For instance, the UK's built environment consumes 25% of raw materials and contributes 62% of total solid waste generation [2]. In the United States, 600 million tonnes of construction debris were generated in 2018, twice the total municipal solid waste generated [3]. The consequences of uncontrolled waste generation resulted in excessive raw material extraction, carbon emissions, energy usage, pollution, and landfilling. The report from [4] highlights that material waste from new constructions contributes to 3.5 million tonnes (MT) of carbon dioxide equivalent (CO_2e) emissions. Moreover, 32% of the UK landfill waste is contributed by waste generated from construction and demolition projects [1]. In addition to environmental aspects, CW significantly impacts project time, cost, and health and safety factors [5–8]. The waste generation from the built environment results from various activities across the supply chain, including design, procurement, construction, operation and maintenance, and demolition. As a result, all stakeholders involved across the supply chain have a role in minimising waste generation and their associated impacts, including the client [9]. However, among all these factors, the literature claims that design-related factors are the leading causes of CW, contributing to over 30% of total waste generation [10]. Besides, it is claimed that waste-driven decisions in the design phases would support the built environment to enhance waste reduction across the project life cycle [11, 12].

The last ten decades of the Industrial Revolution were dominated by a traditional linear economy model, in which the products are purchased, used, and discarded at the end of their life, leading to increased waste generation. However, with the prospects of a surging global population coupled with the changes in consumption patterns, considerable changes to the environment, health, and well-being are becoming integral parts of our system. As an alternative to the linear economy, the circular economy (CE) principles are on the frontline of decision-making to enhance the productive use of CW and maintain their values for an extended period [13–17]. The shift from a traditional fragmented system to a collaborative digital

DOI: 10.1201/9781003610762-1

environment within the built environment, underpinned by Building Information Modelling (BIM), is also seen as a major player in managing CW levels [18–21]. Besides, waste reduction in the built environment is also achieved by integrating modern construction methods, innovation technologies, and design-out waste principles [22]. Apart from the technological developments, the literature emphasises that managing stakeholder-associated factors also contributes to managing CW generation across the life cycle stages [23, 24]. However, on top of all these approaches, accurate quantification and classification of material waste generation from the built environment are a prerequisite for developing solid construction waste management (CWM) plans [25–27]. Thus, measuring the amount, type, time, and place of CW generation is essential for developing efficient management measures. This led to the development of various digital models that support construction waste quantification (CWQ) at the building, regional, and national levels. Also, the CW generation data acts as one of the leading indicators in the life cycle assessment (LCA) models [28].

1.2 Building Information Modelling

BIM is widely applied in the built environment to increase the efficiency of projects across their life cycle. The BIM functionalities, including the life cycle data, simulation and analysis, better design, automated process, and quality documentation, facilitate project scheduling, resource management, cost estimation, facilities management, and sustainability analysis [18, 29]. Building Information Model, the digital model developed using BIM supports the stakeholders in visualising the complete facility that will be built [18]. Besides, the BIM models aid in detecting clashes between various components in the system and managing them in a virtual environment to avoid reworks on-site [20, 30]. These functionalities further improved information sharing, communication, and collaboration among supply chain members. Besides, the application of BIM is also widely recognised in assessing and managing CW workflows (discussed in Section 2.6.1). The life cycle information, interoperability with external platforms, simulation, and visualisation features of BIM support the CWQ across the design, construction, and demolition stages [31]. However, the existing models focus highly on enhancing the quantification process while paying less attention to the semantics, quality, granularity, and standardisation of the CW data (discussed in Section 2.7). As a result, the outputs from these CWQ models lack integration with other decision-making tools, challenging interoperability and automation. Hence, the research seeks to propose a framework with standard semantics, high granularity, and enhanced dynamics with BIM that supports CW assessment from the early design stages.

1.3 Semantic Web Technology

With the advancement of digital technologies, the quantity of data used increases. The piles of data generated from each platform must be seamlessly integrated to derive meanings from the datasets [32]. However, the lack of standardised information and ununified structure hinders data integration across several platforms. Semantic Web Technology (SWT) fills these gaps by allowing the machines to analyse data independently and interact seamlessly with various systems, applications, and disciplines [33]. Besides, ontologies developed using SWT can integrate data from multiple sources, discover patterns, and interpret meaningful insights from the data structure. Ontologies are metadata models that provide consistent vocabularies with standard semantics that ease communication across domains [34]. Since the data required for CWQ depends on multiple data sources, including design, construction, material databases, and CW platforms, ensuring standardisation and consistency is crucial. Thus, integrating SWT functionalities, a CW data model is proposed in this research to ensure that the CW data structure is unified, standardised, and consistent. Furthermore, the integration of BIM and SWT, as evidenced in the literature [6, 28, 35], supported the research to ensure that the overall workflow is interoperable and automated.

1.4 Knowledge Gap and Research Justification

As a result of complex non-linear material flows in the built environment, a massive amount of information needs to be collected, organised, processed, and mapped to make informed judgements. Similarly, the information required to make waste-informed decisions comes from multiple data sources; its direct application into the CWQ models is constrained due to heterogeneity, inconsistency, and non-interoperability. With the advancement in ICTs, earlier research studies adopted BIM and Artificial Intelligence (AI) features to reduce quantification errors and enhance the overall process. However, existing research models neglect to close the gaps related to heterogeneous design, material, and waste information, causing interoperability and automation issues. Besides, there is a paucity of literature to identify and standardise the critical type, quality, and quantity of information required for automating and improving the CWQ process. Although very few studies [36, 37] adopted the coding system to reduce heterogeneity between design and waste data, the mapping models proposed are mechanical, study-specific, and lack wide-scale applications. The absence of standardised and interoperable information to automate the quantification workflows hinders the applications of existing models on a larger scale. Also, this gap generates multiple non-compatible and non-reliable databases acting in silos rather than a single collaborative data model. The literature also suggests that with the growing usage of information and communication technologies (ICTs), it is critical for research studies to ascertain a reliable data representation to reduce the errors caused by inconsistent definitions [38]. Moreover, implementing a robust ICT-integrated information platform is encouraged to retrieve the information required for making data-driven quality and timely CW decisions from the early design stages [16, 39, 40].

Besides, the current studies lean towards weight- and volume-based estimations, so the arguments made from those estimations tend to focus more on inert waste because of its heavy weight and ignore hazardous waste because of its negligible weight and volume [25, 35, 41–44]. These results lead to wrong directions, where some lightweight materials with high environmental impacts are overlooked [35, 45, 46]. Accounting for these problems, the Department of Environment, Food and Rural Affairs (DEFRA) proposed a shift from a weight-based to an impact-based waste metric system and encourages decision-makers to set targets based on CW impact levels [47]. It is also emphasised that this shift would ensure that the waste domain supports net zero carbon targets. Besides, inadequate granularity in the waste results, narrow focus towards a specific life cycle stage, use of more generic waste factors, and unstandardised and ununified data structures were also some of the significant limitations of the existing CWQ models. Thus, maintaining a unified data structure, standardising the information flows, and ensuring consistency in the results that could ease dynamic data flow across the design, construction, and CW platforms will be the potential solution.

Based on these identified knowledge gaps within the current CWQ workflow and future prospects, this research aims to bridge the data gaps by proposing a BIM-integrated semantic framework that supports CWQ from the early design stages. Furthermore, the semantic data model and a BIM-based digital tool developed by applying the proposed framework ensure seamless information exchange, accessibility, data quality, semantic relationships, and standardised workflows across the life-cycle stages. The overall framework development is underpinned by incorporating the functionalities of BIM and SWT. BIM is utilised for its rich life-cycle data repository of design and material details, automated quantity estimation, parametric modelling, visualisation, and application programming interface (API) features. Since SWT is recognised for developing ontological data models using machine-readable language, it expresses the relationships between CW and BIM data. In particular, the linked data feature of SWT facilitates the development of a data model with a standard structure to ease information exchange and communication between digital platforms. Thus, by integrating the linked data functionalities of SWT into BIM, this research ensures interoperability and automation in quantifying CW and informing optimal material choices.

1.5 Research Aim, Objectives, and Scope

The research aims to develop a BIM-integrated semantic framework to quantify CW impacts and inform optimal material choices from the early design stages. This framework aims to achieve standardisation, granularity, interoperability, and automation across the CWQ workflow through the integration of BIM, SWT, and LCA. The research aim is achieved through the following objectives:

1. Understand the overall scenario of CWM methods and their applications across the project life cycle.
2. Explore the functionalities and applications of existing CWQ models.
3. Identify the attributes of BIM, SWT, and LCA for developing a unified and integrated CWQ framework.

4. To establish the industry requirements and identify the critical indicators crucial for making CW-informed decisions.
5. To develop a BIM-integrated semantic CWQ framework accounting for the strategies and industry requirements established in the previous objectives.
6. To demonstrate the application of the proposed framework using an ontology and a BIM-integrated digital tool.
7. To validate the framework against a real-life test case building to demonstrate its adaptability and accuracy.

The scope of the research includes the following:

- The built environment is comprised of two main sub-sectors: buildings and infrastructure. However, the book has considered only the building projects, so the developed framework would be applicable only within this scale.
- The framework proposed in the book is developed to align with the design stages of the life cycle. This is because of the significant impact of design stages on enhancing preventive CWM measures (discussed in Section 2.2.3). According to the Royal Institute of British Architects (RIBA) Plan of Work 2020, the 'design stages' include concept design (RIBA stage 2), spatial coordination (RIBA stage 3), and technical design (RIBA stage 4) stage, where the functionalities of the developed tool can be fully utilised.
- The CW generation significantly impacts the environmental, economic, and social pillars of sustainability (discussed in Section 2.2.4). However, the global warming potential (GWP), one of CW's environmental impacts, is only included in the book.

1.6 Design Science Research

The research follows the six-step flexible approach proposed in the design science research (DSR) framework. Being a problem-solving paradigm, DSR aims to develop an innovative solution to real-world problems by generating innovation artefacts and design knowledge. Such generated design knowledge may be constructs, models, methods, and instantiations [48]. The research outcome aligns with the 'method' knowledge, where the guidelines and workflows are proposed to solve the interoperability and automation gaps within the CW domain. There are six steps associated with the DSR approach [48]. It includes the following:

- Step 1 – problem identification
- Step 2 – define research objectives
- Step 3 – design and development of the solution
- Step 4 – demonstration or application of the solution
- Step 5 – evaluation
- Step 6 – dissemination and communication of the knowledge

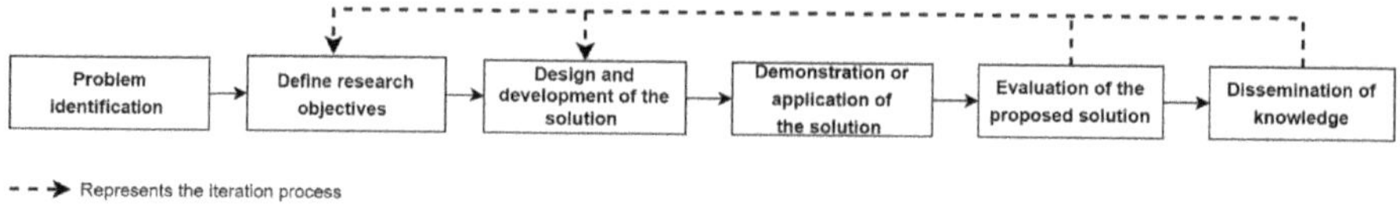

Figure 1.1 DSR Process

Source: Drafted by the author based on descriptions provided in the literature

Figure 1.1 depicts the six steps of the DSR approach. *Steps 1* and *2* (problem identification and defining research objectives) of DSR are discussed in Chapters 1 and 2. The solution proposed in the study (*Step 3*) is a semantic framework that aims to develop an interoperable and automated workflow supporting CWQ from the early design stages of building projects and inform optimal material choices from the obtained waste results. The existing research gaps (identified from the SLR in Chapter 2), the functionalities of BIM, SWT, and LCA (Chapter 3), perspectives of stakeholders engaged across the construction supply chain, and the identification of indicators for CW-driven decision-making (Chapter 4) are integrated to develop the framework. The application of the proposed framework (*Step 4*) is demonstrated via a semantic web application (Chapter 5) and a BIM-integrated application (Chapter 6). These applications led to the development of an ontology and a BIM-enabled user interface (UI), which are tested using the test case building data (*Step 5*). *Step 7* is about the dissemination of the study results.

References

1. WorldGBC, *Beyond the business case – Why you can't afford not to invest in a sustainable built environment*, 2021. https://worldgbc.org/wp-content/uploads/2022/08/WorldGBC-Beyond-the-Business-Case.pdf. Last access: 14 December 2024.
2. Defra, *UK statistics on waste*. 2022. https://www.gov.uk/government/statistics/uk-waste-data/uk-statistics-on-waste. Last access: 04 January 2023.
3. US EPA, *Construction and demolition debris: Material-specific data*, 2021. https://www.epa.gov/facts-and-figures-about-materials-waste-and-recycling/construction-and-demolition-debris-material Last access: 17 May 2022.
4. Construction Leadership Council, *The routemap for zero avoidable waste in construction*, 2021. https://www.constructionleadershipcouncil.co.uk/wp-content/uploads/2021/07/ZAW-Interactive-Routemap-FINAL.pdf. Last access: 05 January 2023.
5. T. Ann, C.S. Poon, A. Wong, R. Yip and L. Jaillon, Impact of construction waste disposal charging scheme on work practices at construction sites in Hong Kong. *Waste Management*, 2013. 33(1): pp. 138–146. https://doi.org/10.1016/j.wasman.2012.09.023
6. S. Nagapan, I.A. Rahman, A. Asmi, A.H. Memon and I. Latif, Issues on construction waste: The need for sustainable waste management, in *2012 IEEE colloquium on humanities, science and engineering (CHUSER)*, 2012. IEEE. https://doi.org/10.1109/CHUSER.2012.6504333
7. M.B. Purushothaman and J. Seadon, System-wide construction waste and their connectivity to construction phases, impacting 5M factors and effects: A systematic review. *Smart and Sustainable Built Environment*, 2024. 13(2): pp. 354–369. https://doi.org/10.1108/SASBE-12-2022-0269

8. J. Sharman, *Construction waste and materials efficiency*, 2018. NBS. https://www.thenbs.com/knowledge/construction-waste-and-materials-efficiency. Last access: 05 August 2022.
9. S. Sivashanmugam, R. Sergio, R. Farzad and D. Nashwan, Maximising the construction waste reduction potential – Barriers and catalysts. *Proceedings of the Institution of Civil Engineers – Civil Engineering*, 2022. 176(6): pp. 1–9. https://doi.org/10.1680/jcien.22.00163
10. M. Osmani, Construction waste minimization in the UK: Current pressures for change and approaches. *Procedia-Social and Behavioral Sciences*, 2012. 40: pp. 37–40. https://doi.org/10.1016/j.sbspro.2012.03.158
11. Z. Liu, M. Osmani, P. Demian and A. Baldwin, A BIM-aided construction waste minimisation framework. *Automation in Construction*, 2015. 59: pp. 1–23. https://doi.org/10.1016/j.autcon.2015.07.020
12. M. Osmani, Design waste mapping: A project life cycle approach, in *Proceedings of the institution of civil engineers-waste and resource management*, 2013. ICE Publishing. https://doi.org/10.1680/warm.13.00013
13. K.T. Adams, M. Osmani, T. Thorpe and J. Thornback, Circular economy in construction: Current awareness, challenges and enablers, in *Proceedings of the institution of civil engineers-waste and resource management*, 2017. Thomas Telford Ltd. https://doi.org/10.1680/jwarm.16.00011
14. P. Ghisellini, C. Cialani and S. Ulgiati, A review on circular economy: The expected transition to a balanced interplay of environmental and economic systems. *Journal of Cleaner Production*, 2016. 114: pp. 11–32. https://doi.org/10.1016/j.jclepro.2015.09.007
15. M.U. Hossain, S.T. Ng, P. Antwi-Afari and B. Amor, Circular economy and the construction industry: Existing trends, challenges and prospective framework for sustainable construction. *Renewable and Sustainable Energy Reviews*, 2020. 130: p. 109948. https://doi.org/10.1016/j.rser.2020.109948
16. Y. Yu, D.M. Yazan, V. Junjan and M.-E. Iacob, Circular economy in the construction industry: A review of decision support tools based on Information & Communication Technologies. *Journal of Cleaner Production*, 2022: p. 131335. https://doi.org/10.1016/j.jclepro.2022.131335
17. C. Zhang, M. Hu, F. Di Maio, B. Sprecher, X. Yang and A. Tukker, An overview of the waste hierarchy framework for analyzing the circularity in construction and demolition waste management in Europe. *Science of the Total Environment*, 2022. 803: p. 149892. https://doi.org/10.1016/j.scitotenv.2021.149892
18. S. Azhar, Building information modeling (BIM): Trends, benefits, risks, and challenges for the AEC industry. *Leadership and Management in Engineering*, 2011. 11(3): pp. 241–252. https://doi.org/10.1061/(ASCE)LM.1943–5630.0000127
19. V.W. Tam, C.M. Tam, S. Zeng and W.C. Ng, Towards adoption of prefabrication in construction. *Building and Environment*, 2007. 42(10): pp. 3642–3654. https://doi.org/10.1016/j.buildenv.2006.10.003
20. J. Won and J.C. Cheng, Identifying potential opportunities of building information modeling for construction and demolition waste management and minimization. *Automation in Construction*, 2017. 79: pp. 3–18. https://doi.org/10.1016/j.autcon.2017.02.002
21. H. Yuan, W. Lu and J.J. Hao, The evolution of construction waste sorting on-site. *Renewable and Sustainable Energy Reviews*, 2013. 20: pp. 483–490. https://doi.org/10.1016/j.jclepro.2012.08.016
22. P. Nie, K.C. Dahanayake and N. Sumanarathna, Exploring UAE's transition towards circular economy through construction and demolition waste management in the preconstruction stage–A case study approach. *Smart and Sustainable Built Environment*, 2024. 13(2): pp. 246–266. https://doi.org/10.1108/SASBE-06-2022-0115
23. E.Y. Frempong-Jnr, S.O. Ametepey and J.E. Cobbina, Impact of stakeholder management on efficient construction waste management. *Smart and Sustainable Built Environment*, 2023. 12(3): pp. 607–634. https://doi.org/10.1108/SASBE-08-2021-0147

24. X. Zhao, Stakeholder-associated factors influencing construction and demolition waste management: A systematic review. *Buildings*, 2021. 11(4): p. 149. https://doi.org/10.3390/buildings11040149
25. A. Bakshan, I. Srour, G. Chehab and M. El-Fadel, A field based methodology for estimating waste generation rates at various stages of construction projects. *Resources, Conservation and Recycling*, 2015. 100: pp. 70–80. https://doi.org/10.1016/j.resconrec.2015.04.002
26. B. Bossink and H. Brouwers, Construction waste: Quantification and source evaluation. *Journal of Construction Engineering and Management*, 1996. 122(1): pp. 55–60. https://doi.org/10.1061/(ASCE)0733–9364(1996)122:1(55)
27. J. Solís-Guzmán, M. Marrero, M.V. Montes-Delgado and A. Ramírez-de-Arellano, A Spanish model for quantification and management of construction waste. *Waste Management*, 2009. 29(9): pp. 2542–2548. https://doi.org/10.1016/j.wasman.2009.05.009
28. S. Sivashanmugam, S. Rodriguez Trejo and F. P. Rahimian, BIM-integrated semantic framework for construction waste quantification and optimisation. *Automation in Construction*, 2024. 168: p. 105842. https://doi.org/10.1016/j.autcon.2024.105842
29. S. Azhar and J. Brown, BIM for sustainability analyses. *International Journal of Construction Education and Research*, 2009. 5(4): pp. 276–292. https://doi.org/10.1080/15578770903355657
30. J. Won, J.C. Cheng and G. Lee, Quantification of construction waste prevented by BIM-based design validation: Case studies in South Korea. *Waste Management*, 2016. 49: pp. 170–180. https://doi.org/10.1016/j.wasman.2015.12.026
31. S. Sivashanmugam, S. Rodriguez, F.P. Rahimian, F. Elghaish and N. Dawood, Enhancing information standards for automated construction waste quantification and classification. *Automation in Construction*, 2023. 152: p. 104898. https://doi.org/10.1016/j.autcon.2023.104898
32. N. Shadbolt, T. Berners-Lee and W. Hall, The semantic web revisited. *IEEE Intelligent Systems*, 2006. 21(3): pp. 96–101. https://doi.org/10.1109/MIS.2006.62
33. T. Berners-Lee, J. Hendler and O. Lassila, The semantic web. *Scientific American*, 2001. pp. 34–43. https://www-sop.inria.fr/acacia/cours/essi2006/Scientific%20American_%20Feature%20Article_%20The%20Semantic%20Web_%20May%202001.pdf. Last access: 02 March 2023.
34. N.F. Noy and D.L. McGuinness, *Ontology development 101: A guide to creating your first ontology*, 2001. Stanford Knowledge Systems Laboratory. https://protege.stanford.edu/publications/ontology_development/ontology101.pdf. Last access: 08 December 2024
35. P. Sun, N. Zhang, J. Zuo, R. Mao, X. Gao and H. Duan, Characterizing the generation and flows of building interior decoration and renovation waste: A case study in Shenzhen City. *Journal of Cleaner Production*, 2020. 260: p. 121077. https://doi.org/10.1016/j.jclepro.2020.121077
36. R. Quiñones, C. Llatas, M.V. Montes and I. Cortés, A multiplatform BIM-integrated construction waste quantification model during design phase. The case of the structural system in a Spanish building. *Recycling*, 2021. 6(3): p. 62. https://doi.org/10.3390/recycling6030062
37. R. Quiñones, C. Llatas, M.V. Montes and I. Cortés, Quantification of construction waste in early design stages using BIM-based tool. *Recycling*, 2022. 7(5): p. 63. https://doi.org/10.3390/recycling7050063
38. V.W. Tam, Y. Zhou, C. Illankoon and K.N. Le, A critical review on BIM and LCA integration using the ISO 14040 framework. *Building and Environment*, 2022. 213: p. 108865. https://doi.org/10.1016/j.buildenv.2022.108865
39. L.A. Akanbi, A.O. Oyedele, L.O. Oyedele and R.O. Salami, Deep learning model for demolition waste prediction in a circular economy. *Journal of Cleaner Production*, 2020. 274: p. 122843. https://doi.org/10.1016/j.jclepro.2020.122843

40. L.A. Akanbi, L.O. Oyedele, O.O. Akinade, A.O. Ajayi, M.D. Delgado, M. Bilal and S.A. Bello, Salvaging building materials in a circular economy: A BIM-based whole-life performance estimator. *Resources, Conservation and Recycling*, 2018. 129: pp. 175–186. https://doi.org/10.1016/j.resconrec.2017.10.026
41. F. Kleemann, H. Lehner, A. Szczypińska, J. Lederer and J. Fellner, Using change detection data to assess amount and composition of demolition waste from buildings in Vienna. *Resources, Conservation and Recycling*, 2017. 123: pp. 37–46. https://doi.org/10.1016/j.resconrec.2016.06.010
42. J. Li, Z. Ding, X. Mi and J. Wang, A model for estimating construction waste generation index for building project in China. *Resources, Conservation and Recycling*, 2013. 74: pp. 20–26. https://doi.org/10.1016/j.resconrec.2013.02.015
43. A. Vilventhan, V. Ram and S. Sugumaran, Value stream mapping for identification and assessment of material waste in construction: A case study. *Waste Management & Research*, 2019. 37(8): pp. 815–825. https://doi.org/10.1177/0734242X19855429
44. H. Wu, H. Duan, L. Zheng, J. Wang, Y. Niu and G. Zhang, Demolition waste generation and recycling potentials in a rapidly developing flagship megacity of South China: Prospective scenarios and implications. *Construction and Building Materials*, 2016. 113: pp. 1007–1016. https://doi.org/10.1016/j.conbuildmat.2016.03.130
45. J. Wang, H. Wu, H. Duan, G. Zillante, J. Zuo and H. Yuan, Combining life cycle assessment and building information modelling to account for carbon emission of building demolition waste: A case study. *Journal of Cleaner Production*, 2018. 172: pp. 3154–3166. https://doi.org/10.1016/j.jclepro.2017.11.087
46. J.P. Xu, Y. Shi, Y.C. Xie and S.W. Zhao, A BIM-based construction and demolition waste information management system for greenhouse gas quantification and reduction. *Journal of Cleaner Production*, 2019. 229: pp. 308–324. https://doi.org/10.1016/j.jclepro.2019.04.158
47. Defra, *Our waste, our resources: A strategy for England*, 2018. https://assets.publishing.service.gov.uk/government/uploads/system/uploads/attachment_data/file/765914/resources-waste-strategy-dec–2018.pdf. Last access: 02 April 2023.
48. J. Vom Brocke, A. Hevner and A. Maedche, Introduction to design science research, in *Design science research. Cases*, 2020. Springer, pp. 1–13. ISBN: 978-3-030-46781-4.

2 Construction Waste – Definition, Classification, Sources Impacts, Management, and Quantification

2.1 Introduction

An increase in built environment activities significantly increases the CW generation. As a result of a complex and longer supply chain, the waste generated from the built environment is distributed, non-linear, and heterogeneous. Excessive waste generation has detrimental impacts on environmental, economic, and social sustainability. Various innovative technologies are introduced to enhance how waste is managed across the construction supply chain. Incorporating CE in decision-making also supports the built environment in managing waste efficiently from the early design stages to the end of life (EoL). Besides, the regulatory frameworks and legal policies support waste reduction and management across the project life cycle. Of all existing approaches, it was evident from the literature that accurate quantification and classification of CW are prerequisites for their effective management across the project life cycle. A list of 35 CWQ models was identified from the literature. They are grouped into five categories based on the type of ICT used in the models. This includes BIM-integrated CWQ frameworks, AI-based CW prediction and classification models, on-site CW estimation using robots, Bigdata-enabled waste generation rate (WGR) benchmarking, and geographical information system (GIS)-integrated CW models. The benefits identified by integrating ICTs in CWQ are that it minimises the level of human intervention in the CW assessment, increases the accuracy of quantification results, and reduces the time and cost associated with manual processes. Despite these advantages, there are some predominant gaps that limit the wide-scale applications of existing CWQ models. It includes the lack of unified and standardised data structure, focus on unidirectional communication with BIM, inadequate assessment of CW impacts, lack of granularity in the CWQ results, and lack of focus on the whole life cycle while conducting CWQ, less focus on refurbishment and renovation activities, and limited attention on infrastructure and civil engineering works. The applications of each model, as well as their benefits and limitations in developing an information standard and collaborative CWQ workflow, are discussed in this chapter. Overall, the literature review provides the current scenario of CWM, CWQ, and existing research gaps. Besides, the chapter highlights the functionalities that are vital to

DOI: 10.1201/9781003610762-2

be integrated into future CWQ models to streamline the CW assessment, management, and reporting workflows.

2.2 Construction Waste

Over 40–50% of the total solid waste generated in the last decade is an output of construction, demolition, and excavation activities [1]. In the UK, 137.8 million tonnes of solid waste were generated from the construction industry in 2018. This amount contributes to 62% of total solid waste generation in the UK. Although the recent report [2] states that around 92% of non-hazardous CW has recovered, the waste generated at first instance remains high [3]. The literature highlights that the built environment considers CW an inevitable by-product [4]. This section briefly describes how the built environment defines CW, the types of CW classification approaches, sources of CW generation across various supply chain activities, and impacts across the triple bottom line of sustainability. Furthermore, the existing CWM and CWQ approaches and the level of integration of ICTs in the CWQ process are discussed. Finally, the section discusses the significant limitations of existing CWQ models that affect their wide-scale implementation in the built environment.

2.2.1 Definition of CW

Waste Framework Directive (WFD) defines waste as 'any substance or object which the holder discards or intends or is required to discard' [5] – P9. The material waste from the built environment may be a result of several activities during the construction, operation, maintenance, renovation, and demolition of any new and existing buildings and civil engineering structures. Civil engineering structures include public projects such as roads, bridges, tunnels, water, and irrigation systems [6]. However, the literature [7] states that the terms 'Construction Waste' and 'Construction and Demolition Waste' are broadly used to represent the material waste generation from the built environment. So, the term 'Construction Waste' used in the book corresponds to the material waste generated from built environment activities.

2.2.2 Types of CW Classification

Unlike other industrial waste materials, CW generation does not occur at specific manufacturing facilities. The generation of CW is distributed and occurs due to various supply chain activities across the life-cycle stages. This distributed generation of CW causes significant challenges in efficiently managing them. Hence, classifying and reporting CW are vital to enhance the efficiency of CWM. There is no strict approach to classifying CW, so existing studies have adopted various methods [8]. Around eight classification approaches broadly adopted in the literature to report the CW generation are included for discussion.

(i) **The reactive nature of waste materials** – Classifying CW based on its reactive nature involves evaluating how materials may react chemically or physically, especially accounting for their environmental impacts. The three main categories under this classification include inert, non-inert, and hazardous waste. The *inert waste* does not undergo any significant physical or chemical change in the environment. They are generally stable and pose minimal risk to the environment. This includes bricks, masonry blocks, concrete, glass, plaster, sand, and uncontaminated soil. Inert waste can often be recycled or reused in construction applications, such as backfilling, road base, or aggregate for new concrete. *Non-inert waste* can undergo any significant physical or chemical change in the environment. Gypsum, wood, insulation, packaging, and specific plastics are classified under this category. Non-inert waste requires proper management or disposal to mitigate environmental impacts. The *hazardous waste* contains harmful substances that pose significant threats to the environment and human health. The materials include asbestos, paints and varnishes, chemical solvents, and contaminated soil. Hazardous waste requires specialised handling and disposal and should adhere to legal waste management (WM) laws.

(ii) **Material types** – The classification is based on the material composition of CW generated. This approach enables more effective WM strategies, including reuse, recycling, and disposal. Bricks/blocks, concrete, mortar, and metals are considered large contributors to CW [9–12].

(iii) **Waste characterisation codes** – A unique six-digit code called the European Waste Catalogue (EWC) is used in the EU to characterise each waste material [13]. This is called List of Waste (LoW) in the UK. These individual waste materials are grouped as (i) insulation and asbestos materials, (ii) concrete, bricks, tiles, and ceramics, (iii) wood, glass, and plastic, (iv) bituminous mixtures, coal tar and tar, (v) metallic waste, including cable, (vi) soil, contaminated soils, stones, and dredging spoil, (vii) gypsum, (viii) cement, (ix) paints and varnishes, and (x) adhesives and sealants. For example, the brick waste from the built environment is marked as 17–01–02. The first tier represents the source of waste, the second tier represents the material group, and the third tier represents the individual material. In the case of brick, '17' represents the source as the built environment, '01' represents the material group 'concrete, bricks, tiles, and ceramics', and '02' represents the brick. These codes are divided further into four types based on their hazardous nature. This includes 'absolute non-hazardous', 'mirror non-hazardous', 'mirror hazardous', and 'absolute hazardous'. The first and fourth divisions show that the waste will always be non-hazardous and hazardous, respectively. The second and third divisions show that the waste may be non-hazardous or hazardous [14]. These characteristics and the codes reflect the exact state of each material type to allow for a more comprehensive and effective WM. Code selection depends on the division at which the generated waste falls.

(iv) **Life-cycle stages** – The project life cycle stages, including construction, refurbishment and renovation, and demolition, are the main categories adopted in the literature [8]. These stages are widely used to provide a broad picture of waste generation from the built environment. Of all the life cycle stages, it is estimated that demolition activities contribute to 90% of total waste generation, and less than 10% is a result of construction activities [6]. There are no quantitative reports on waste generation during the operation and maintenance stages.

(v) **Origin and characteristics of waste at each life-cycle stage** – The two main categories here are avoidable and non-avoidable (unavoidable) waste [15]. The *avoidable waste* occurs due to ineffective planning, design, and management. Design errors, over-ordering, non-standardised designs, lack of comprehensive WM plans, lack of collaboration and communication, and limited early-stage decisions and inefficiencies are some of the leading causes of this waste. Effective planning, design, management, and training can prevent avoidable waste generation. *Non-avoidable waste* naturally occurs as part of the construction process, and it is challenging to eliminate it entirely, regardless of effective WM practices. This includes breakage, materials that have no reusability, and natural offcuts [16].

(vi) **Activities associated with construction, maintenance, and demolition of buildings** – This method broadly classifies waste into structural and finishing waste. Structural waste is generated from the activities associated with the construction, maintenance, and demolition of the structural components of the sub-structure and super-structure of a building. The material composition of structural waste mainly includes bricks, concrete, masonry, metals, timbers, tiles, and ceramics. Meanwhile, finishing waste arises from the activities incorporated into the project to increase the functionality and aesthetics of the building. The plastering, painting, flooring materials, fixtures and fittings, and glass and packaging materials are dominant in this waste type [17].

(vii) **Origin of material sources** – The origin and composition of materials in the generated waste are used to facilitate classification. The three main categories are organic, inorganic, metallic, and composite waste [8]. Hazardous waste is another primary waste type under this classification [18].

(viii) **Waste management routes** – The type of treatment used for managing CW is used for classification. The three main categories under this approach are reusable, recyclable, and landfillable waste. Few studies, like [19], included this in reporting CW generation.

These are some CW classification approaches adopted in the literature and legal standards. The choice of classification depends on why the CW data is collected, where it will be used, and who the end-users/beneficiaries are. So, selecting an approach that best meets the requirements is vital. In this book, the material-based classification is adopted to classify waste. It is also mapped with EWC codes to ease reporting and align with the legal requirements in the UK.

2.2.3 Sources of CW

CW occurs from various activities across the supply chain, influenced directly or indirectly by the stakeholders involved in project execution [20, 21]. Some of the significant CW-contributing activities discussed in the literature are grouped according to the activities across the project life cycle stages. Table 2.1 shows the factors associated with design, procurement, construction, material handling and storage, management, site operations, and residual activities.

Around one-third – 33% of CW generation – is influenced by design factors and decisions made in the design stages [22]. Specifically, the waste generation caused by design documentation, errors, and reworks is significantly high [23, 24]. In addition, the technical competencies of designers, dimensional coordination, optimised layouts, design standardisation, and error-free and detailed design specifications are prerequisites to encourage the buildability of designs [25]. Improved buildability and managing resource efficiency are proven ways to minimise CW generation [26, 27]. Meanwhile, attitudes and involvement of design team members influence waste generation levels. For instance, by identifying the impacts of waste sources that align with RIBA stages, Osmani [21] emphasises that architects perceive CW generation as a result of errors from on-site operations and that it is more unlikely to be related to design attributes.

Despite the individual perspectives, the literature firmly suggests that design factors, design phase, and design team members are significant influencers of waste reduction across the project execution stages. Thus, design stages are assumed to have a higher probability of reducing waste generation than the later stages of a project life cycle [37]. Besides, the widely established MacLeamy curve recognises the potential of design stages for creating high impacts on project outcomes at lower costs. Also, the cost of changes made in the design stage is relatively economical compared to the later stages [38]. The same scenario applies to the CWM: adopting changes in the virtual design environment will significantly minimise waste generation levels and strengthen WM measures at lower costs. So, the research aims to propose a framework that supports designers in measuring CW generation for their designs, making design changes, and selecting optimal material choices in a virtual environment before on-site construction work begins.

2.2.4 CW Impacts

Excessive generation of CW could have detrimental effects on the built environment's goal of sustainability. The literature highlights the impacts CW causes in both short and long terms across the triple bottom line of sustainability: environmental, economic, and social. This section discusses these impacts.

2.2.4.1 Environmental Impacts

Two-thirds of the total solid waste generated in the UK results from construction, demolition, and excavation activities [2]. The increasing volume of CW significantly increases raw material consumption, energy usage, greenhouse gas

Table 2.1 Project Activities Causing CW Generation

Factors affecting CW generation			*References*
Design		• Frequent design changes • Complexity in designs • Imprecise and inadequate design specification • Last-minute client requirements • Limited focus on integrating CW dimensions into the design • Lack of knowledge about material standards and alternate choices in the market • Late involvement of contractors • Absence of precise CWM targets and goals • Inadequate coordination and communication between the design and construction team • Delays resulting from slow revisions and distribution of designs among various specialists • Offcuts arising from non-standardised designs • Compromised communication between design and CWM tools	[3, 22, 25, 28–34]
Procurement		• Ordering errors • Shipping errors • Waiting times for replacements • Selection of low-quality materials • Excess packaging	[25, 31, 35]
Construction	Material handling and storage	• Improper material handling • Inappropriate space on-site for safe storage of materials • Unwanted transportation of materials on-site • Longer material storage time • Materials supplied in loose form	[27, 31, 32]
	Management	• Reworks • Deficient site WM plans • Lack of training and awareness of CWM • Unclear CWM roles and responsibilities • Poor information quality • Lack of regulatory frameworks • Lack of digital tools	[3, 22, 27, 28, 32, 36]
	Site operations	• Leftover materials • Unprepared equipment malfunctions • Poor workmanship • Time and cost constraints • Poor coordination and communication among the team members	[3, 22, 31, 32]
Other external factors		• Accidents occurring on-site • Weather, vandalism, theft, and other political factors	

Source: Drafted by the author

emissions, pollution, and landfilling. For instance, the Construction Leadership Council (CLC) reports that material waste from new construction projects emits 3.3 million tCO_2e/year [39]. This includes waste arising from waste materials, transport, and treatment processes. It is also estimated that the process associated with CWM solely contributes to 11% of the total life-cycle carbon emissions from a project [40].

Besides, the increasing CW volume also increases the resource consumption levels, leading to material scarcity and energy demands. It is emphasised that the built environment consumes over 50% of the total raw materials extracted globally. For instance, in England, non-metallic minerals, including sand and gravel, contributed to almost half (451 MT) of the total material footprint (902 MT) in 2019 [38]. In addition, the energy consumed for transporting and treating CW is estimated to be significantly higher [41]. In line with the output flows, 32% of total landfill waste comes from construction and demolition activities, in which 13% of the purchased products are directly sent to landfill sites without being used [37]. Though the recovery rates of CW are increasing, nearly 10–12% of it is still disposed of in landfill sites [2]. Other associated impacts rooted in CW include pollution, land occupancy, contamination, and ecosystem destruction [42].

2.2.4.2 *Economic Impacts*

In addition to the environmental impacts caused by CW, its impact on the economic pillar is highly intense. CW generation directly impacts rising raw material prices, waste transportation and treatment costs, landfill disposal charges and taxes, and project time and cost overruns [20, 31, 43–47]. For instance, the Building Research Establishment (BRE) highlights that the cost of waste generated from the construction of new houses in the UK approximates £1.2 billion/year, which includes the cost of wasted materials (81.5%), labour (12.1%) and skips hired (6.4%) [48]. Even after a decade, a recent update from [39] states that material waste from new constructions costs £11 billion per annum to the UK construction industry.

In addition to material costs, waste disposal costs are constantly increasing. For instance, the standard landfill tax rates in the UK increased from £14/tonne in 2003 to £102.10/tonne in 2023 [49], thus indicating the economic cost of waste. It is also emphasised that project cost overrun associated with CW generation, transportation, and treatment is becoming a significant concern for built environment professionals [46]. For instance, the actual cost of CW, including material, transport, and treatment in the UK, is estimated on average as £1619 per 8-yard skip [50].

2.2.4.3 *Social Impacts*

Similar to rising concerns over CW impacts on environmental and economic pillars, the social dimension has also been significantly affected. Accounting for various indicators, literature [42, 51–53] highlights the following social impacts of CW: public health and safety risks, accidents on-site, use of public space, air pollution, decrease in land values adjacent to landfill sites, relationship, and reputability issues. Unlike the environment and economic pillars, the literature focusing on social dimensions is noticeably low. This might be because social factors are non-amenable to empirical evidence [54]. The indicators used for assessing social factors are more likely qualitative and depend on the knowledge and understanding of stakeholders active across the Architecture, Engineering, Construction, and Operation (AECO) supply chain [55], thus challenging the CWQ process.

2.3 Construction Waste Management

CWM is the process of managing the material waste generated across the project life cycle through various waste reduction and treatment measures. Several techniques, digital tools, process changes, regulatory frameworks, and legal policies are adopted to manage CW generation. Besides, the shift from traditional fragmented construction to a collaborative and digital project environment ensures CW reduction across the life-cycle stages [46, 57–59]. The role of existing approaches and their impacts on CWM is discussed in the following sections.

2.3.1 Technological Developments

Advanced digital technologies are widely applied to overcome the challenges associated with existing CWM practices. BIM, an emerging technology, has become an essential solution for managing CW waste for decades. Despite being labelled as technology, process, and policy, the widely practised BIM features in CWM include 3D visualisation, site utilisation, phase planning (4D BIM), design validation or clash detection, and BIM-based material take-off (MTO). These applications aim to facilitate the supply chain in mitigating the factors that cause waste generation. For instance, BIM-based visualisation and design validation manage the reworks resulting from design errors and complexities being identified as one of the major causes of on-site waste (discussed in Section 2.2.3). Won, Cheng [34] identified that detecting design errors using BIM could prevent a maximum of 15% of CW generation, which otherwise would contribute to construction and installation waste. Besides, the comprehensive information in the BIM models facilitates waste reduction through accurate material ordering, cutting, and layout. For instance, Porwal and Hewage [60] demonstrated that BIM reduced rebar waste by 1.6%. Khondoker [61] proved that BIM-based quantity details have a massive impact on material optimisation while minimising offcut waste from cutting rebars. Besides, BIM is also integrated with other ICTs to enhance CWM across the life-cycle stages. For example, Akinade and Oyedele [62] integrated BIM and AI to propose a CW analytics system that supports CW prediction in the early design stages. Another study by Liu et al. [63] integrated the blockchain functionalities with BIM and City Information Modelling to enhance the CW information flow and management across the life cycle stages.

The literature also claims that the applications of other technologies, including AI, GIS, augmented reality (AR), and virtual reality (VR), promote CW reduction by enhancing the construction, management, and waste treatment processes. For instance, GIS and BIM were integrated to design waste transportation routes by selecting appropriate waste treatment and landfill sites [64]. Machine-learning (ML) and deep-learning (DL) models using AI algorithms play a significant role in measuring, classifying, managing, and optimising CW generation. Besides, a hybrid approach that integrates two or more algorithms or technologies can also be seen in CW research. For example, AI is combined with augmented reality to develop a 3D model from traditional 2D drawings that supports the built environment in developing pre-demolition audit reports for EoL management of the buildings [65]. This study emphasises the automation and accuracy of information as the main benefits of AI-AR integration. Other digital technologies like blockchain, digital twins, big data analytics, and the internet of things also facilitate the built environment to manage the design factors leading to CW generation [66].

2.3.2 Process Changes

Realising the impacts of CW generation, significant changes have been brought into play across the project life cycle to enhance the efficiency of CWM. The process changes are associated with designing, estimation, procurement, project management, and human resources. It is an integrated approach that includes all the supply-chain players: clients, designers, contractors, suppliers, and other stakeholders dealing with CWM. The changes in the WM process are influenced mainly by the advancement of technologies and policies. Since the design stages are considered critical for CW reduction, various measures were implemented to manage the design factors causing waste. The Waste and Resources Action Programme also introduced the Design Out Waste principles to assist the architects and designers in integrating the CW dimension into the design process. The five main principles, design for material optimisation, design for waste-efficient procurement, design for material reuse and recovery, design for off-site construction, and design for deconstruction and flexibility, were proposed to minimise CW generation [67].

Several on-site CWM strategies are practised in the built environment, including on-site waste segregation, prefabrication, supplier take-back schemes, lean construction principles, and reverse logistics. For instance, 52% on-site waste reduction could be achieved by shifting from conventional to prefabricated construction [68], while [69] confirms that the difference in WGR between traditional and prefabrication buildings is estimated at 15 tonnes/m^2. The idea of procuring secondary materials is witnessing a significant impact on WM. For instance, 2500 tonnes of second-hand pipeline tubes were used for roof construction of the 2012 London Olympic Stadium, contributing to 20% of the total steel consumed for roof truss assembly [70]. Besides, [71] emphasises that lean construction practices could potentially reduce on-site waste.

All these process changes applied across the supply chain are more likely to align with the CE principles. CE is broadly presented as a transformative model aiming to decouple resource consumption from economic growth, promoting sustainable development [72, 73]. The CE principles are gaining interest among researchers, business organisations, and governments worldwide because of their operational attributes that help implement sustainable development [72, 74, 75]. It promotes a shift from a traditional 'take-make-dispose' model to a circular one by maximising reuse and recovery rates and eliminating waste generation [76]. The circular model shifts the mindset of waste and waste treatment, which is called 'non-valued added' to a 'valuable resource and process'. The R principles are widely posed as an approach/strategy to implement CE principles across the supply chain. Starting with three primary actions in the 3R (reduce, reuse, and recycle) principle, scholars have proposed 4R, 5R [77], 6R [78, 79], 9R [80, 81], and 10R [82] principles to accelerate CE adoption across the supply chain. The R principles include refuse, rethink, reduce, reuse, repair, refurbish, remanufacture, repurpose, recycle, and recover. The application of CE through various decision-making support systems, evaluation tools, and material banks to mitigate the sustainable impacts associated with excessive raw resource consumption and excessive CW generation is broadly focused.

2.3.3 Policy Implications

The increasing demand for efficient management of resources and the prevention of the negative impacts of CW generation puts intense pressure on organisations, researchers, and the government to develop clear CWM targets and policies [22]. As a result, policymakers introduce various legislative measures, voluntary targets, frameworks, and taxes at project and national levels. Among various regulations, the WFD (2008/98/EC) plays a vital role in measuring and managing CW. It requires waste to be managed without harming the environment, endangering human health, and preventing soil, noise, and water pollution. This directive puts a foundation for the framework, called 'Waste Hierarchy', which proposes an order of preference for its member states to manage waste. The priority is given to 'prevention'. This is followed by 'preparing for reuse', 'recycling', 'recovery', and 'disposal' of waste [5]. Since 'Prevention' focuses on reducing the amount of waste generated from the source, it is also termed 'waste reduction' or 'source reduction or waste minimisation' [37]. The preventive WM approach of the WFD is incorporated in the UK through 'The Waste Regulations (England and Wales) 2011'. The WFD also introduced the 'polluter pays principle' and 'extended producer responsibility', accounting for the fact that the person/organisation/waste holders responsible for waste generation should bear the cost of its management. Furthermore, the recent update in the WFD (2018/851/EC) strengthen the commitments to waste prevention and emphasises the shift from 'WM' to 'material management' to lower raw material consumption and enhance the use of secondary materials [83], which is also the prime focus of CE principles (Figure 2.1). The difference between the WFDs Waste Hierarchy and CE is shown (Table 2.2).

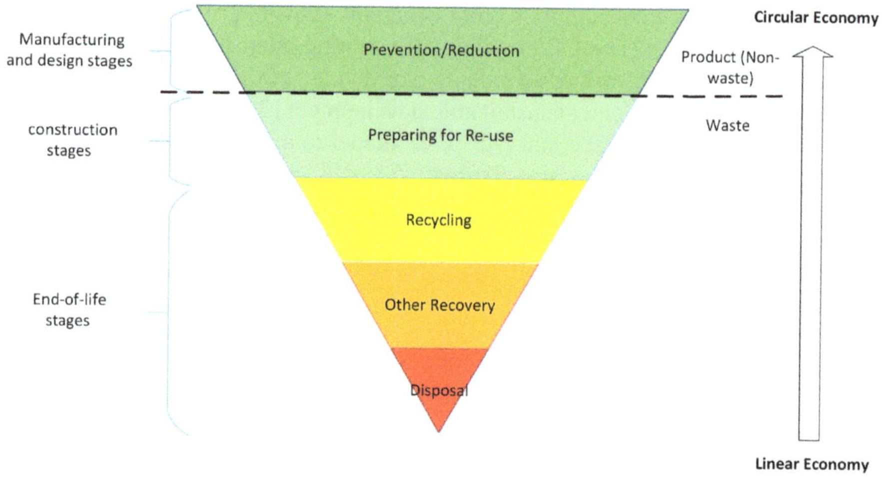

Figure 2.1 Waste Hierarchy and CE

Source: Drafted by the author based on the descriptions given in the literature

Table 2.2 Comparison of Waste Hierarchy and CE

Waste Hierarchy (WFD)	*CE (Ellen MacArthur)*
WFD laid out a framework to propose a set of principles for managing waste. The possible WM methods are selected and prioritised based on their environmental impact levels.	CE is an industrial system that is restorative and regenerative by intention and design. The system aims to rely on renewable energy, eliminate waste through careful design of materials, products, and systems, and reduce and eliminate the use of toxic chemicals [76].
The *Waste Hierarchy* focuses on the output flows (waste).	Both input and output flows (material/energy/land/water) are prioritised in the CE.
The methods proposed in the Waste Hierarchy are likely to align with the waste perspective.	The CE principles are developed from the resource perspective.
Though landfilling of waste is the least preferred method, it is accepted as a WM method.	In the CE, landfilling is unacceptable, and doing so makes the WM a non-circular (linear) economy approach.
The focus on the construction and post-construction phases is high in the framework.	The early design stages are the leading players in CE and emphasise that the whole life-cycle perspective is essential to minimising the impacts.
This framework can be called a strategic approach that supports CE adoption.	CE is a principle that incorporates most of the WM methods proposed in the Waste Hierarchy.

Source: Drafted by the author based on the descriptions given in the literature

The Site Waste Management Plans Regulations 2008 in the UK placed obligations on the contractors and the client to develop a management plan, including the waste expected to be generated from the construction process and the WM and disposal methods likely to be used to treat them. This applies only to projects with an estimated cost of more than £3,00,000; however, the regulations have not been mandatory since 2013. The European Commission Decision 2000/532/EC1 also introduced EWC to report waste. The EWC classifies waste according to activities generated and its material composition (discussed in Section 2.2.2). Besides, including EWCs in the Duty of Care waste transfer and hazardous waste consignment notes is a legal requirement (Environmental Agency, 2012). Other fiscal policies, including landfill taxes, penalties, and disposal charges, have been introduced to drive waste away from landfills. The UK standard and lower landfill tax rate, starting at the rate of £2 and £7 in 1996, has now increased to £102.10 and £3.25 per tonne in 2023 [49]. These WM policies and legislation promote efficient management and transparent reporting of CW.

2.4 Construction Waste Quantification

As a result of an increased material diversity across the life cycle stages of buildings, a detailed evaluation of CW quantity, type, source, time, and impacts becomes essential to select the most suitable treatment routes and allow safe disposal [20, 84, 85]. Thus, evaluating the generation rates, treatment routes, and associated sustainable impacts of CW is termed 'CWQ'. It is widely acknowledged as a prerequisite for effective WM throughout the project life cycle [86, 87]. The CWQ supports the built environment in planning effective WM strategies, setting clear benchmarks and targets, ensuring regulatory compliance, and achieving sustainability targets. Also, the WM steps proposed in the ISO 14001:2015 prioritise the measurement process to store, treat, and dispose of waste materials effectively [88]. Some of the widely adopted methods to quantify CW include material flow analysis (MFA), volumetric measurements obtained from direct site visits, use of WGRs, LCA approach, bill of quantities estimation, mass balance principle, and building stock models [87]. The type of CWQ method depends on the scope of the estimation process and the quality and quantity of CW data available.

Two main terms are used to differentiate the scope of the CWQ process. If the quantification process is performed for a specific project, it is termed project-level quantification. In comparison, evaluating material waste generated across a specific region encompassing multiple projects is termed regional-level quantification [87]. Each of these quantification types has unique applications on CWM. For instance, project-level quantification aims to support comprehensive WM planning across the life-cycle stages, determine the generation rates, and support benchmarking based on the project functionalities. Meanwhile, regional-level quantification supports the government and policymakers in setting legal waste targets and developing WM regulations based on geographical scope [87].

The research supports the project-level quantification process. A CWQ data model applicable in the design stages of a building project is proposed, thus characterising the building-level quantification. Furthermore, utilising the acquired CWQ results, the material deemed less waste-generating and less impactful is optimised.

2.5 Existing CWQ Models

Owing to the increased importance of waste generation and management in the built environment, several digital CWQ models have been developed in the past decade. This book section discusses the methodology adopted to retrieve the existing studies that focus on quantifying CW generation across the life cycle stages. The type of ICTs integrated into CWQ workflows, their key functionalities, and their limitations are discussed in the following sections.

2.5.1 Systematic Literature Review

A systematic literature review (SLR) is adopted in this study to compare and interpret patterns of ICT functionalities integrated into the CWQ workflows. SLRs are advanced from traditional narrative reviews because of the transparent methodology adopted to collect and analyse data [89, 90]. The critical aspect of SLR is that it allows researchers to filter articles based on pre-defined selection criteria, which reduces the ambiguities associated with data handling [91]. It can be defined as a systematic approach used to identify, evaluate, and synthesise research findings that aid in informed decision-making [92]. The five main elements of SLR that need to be defined before the review [91, 93] are tabulated (Table 2.3).

2.5.2 Identifying Relevant Articles

The articles were retrieved from widely accepted academic databases, Web of Science and Scopus [85, 94]. In addition, the international journals with Q1 and Q2 Scopus indexes focusing on construction WM and ICT were added for further retrieval. The journals include *Automation in Construction, Construction and Building Materials, Journal of Cleaner Production, Resources, Conservation and Recycling, Sustainability, Waste Management*, and *Waste Management and Research.* A bibliometric search method using keywords suggested by [95] was adopted to maximise the quality of search results. This method is adopted to collect relevant articles for review. The waste generated from the construction sector may be termed ‘construction waste’, ‘C&D waste’, and ‘demolition waste’. Moreover, quantification includes ‘estimation’, ‘prediction’, ‘forecasting’, and ‘assessment’. Similarly, waste’ classification’ includes ‘categorisation’, ‘sorting’, and ‘composition’. The keywords, including ‘Building Information Modelling’, ‘BIM’, ‘Artificial Intelligence’, ‘AI’, ‘Machine Learning’, ‘ML’, ‘Deep Learning’, ‘DL’, ‘Augmented Reality’, ‘AR’, ‘Virtual Reality, ‘VR’, and ‘Robots’, were added to the search to search the integration of ICT in CWQ. Therefore, the keywords were combined using the Boolean operators OR and (AND) to retrieve articles relevant to the review. For example, ‘construction waste’ OR ‘C&D waste’ OR ‘demolition waste’ AND ‘quantification’ OR ‘estimation’ OR ‘prediction’ OR ‘forecasting’ OR

Table 2.3 SLR Elements and Their Applications in This Study

Phases	*Review elements*	*Applications in the current study*
i)	Framing questions for review	a) What is the role of ICT in CWQ? b) What significant limitations are associated with the ICT-integrated CWQ workflows? c) What functionalities within ICT-integrated CWQ models are vital to reinforce the quantification efficiency?
ii)	Identifying relevant papers – search strategies	a) Source: Electronic databases (a detailed description of the type of database is given below) b) Selected keyword search using Boolean operators (Section 2.5.2)
iii)	Inclusion and exclusion criteria	a) Type – Research articles b) Language – English c) Year of publication – From 2012 to 2024 d) Only papers with empirical evidence are included for review d) Papers that focus on solid construction waste are included, excluding liquid waste, general industrial waste, and municipal waste
iv)	Assessing the quality of papers – screening and eligibility	a) Duplicate screening b) Title, abstract, and keywords were checked to identify the relevance of the paper with the scope of the study
v)	Results and discussion (summarising the data evidence and interpreting the findings)	The data from the retrieved articles are extracted and analysed using NVivo. The qualitative description of the applications of various ICT functionalities in enhancing the CWQ process (Section 2.6), the limitations of existing ICT-integrated CWQ workflows and the future research directions (Section 2.7) are discussed.

Source: Drafted by the author

'assessment' AND 'BIM'. The keywords were used to conduct the searches within the article title. The data collection and screening process adopted to retrieve articles for review are illustrated (Figure 2.2).

2.5.3 Screening and Eligibility

A two-step screening process was adopted to validate the eligibility of retrieved articles. The initial search within the article title using the keywords yielded 167 articles. Then, 81 duplicate articles were removed using the automated duplicate screening option in the 'Endnote'. The remaining 86 articles were retrieved for analysis. In the second stage of screening, the objectives of the articles were confirmed by scanning the title, abstract, and keywords manually. The articles with empirical evidence on construction waste quantities, generation rates, and material composition were only accounted for review. In addition, the articles discussing WM and associated innovations with limited or no numerical CWQ results

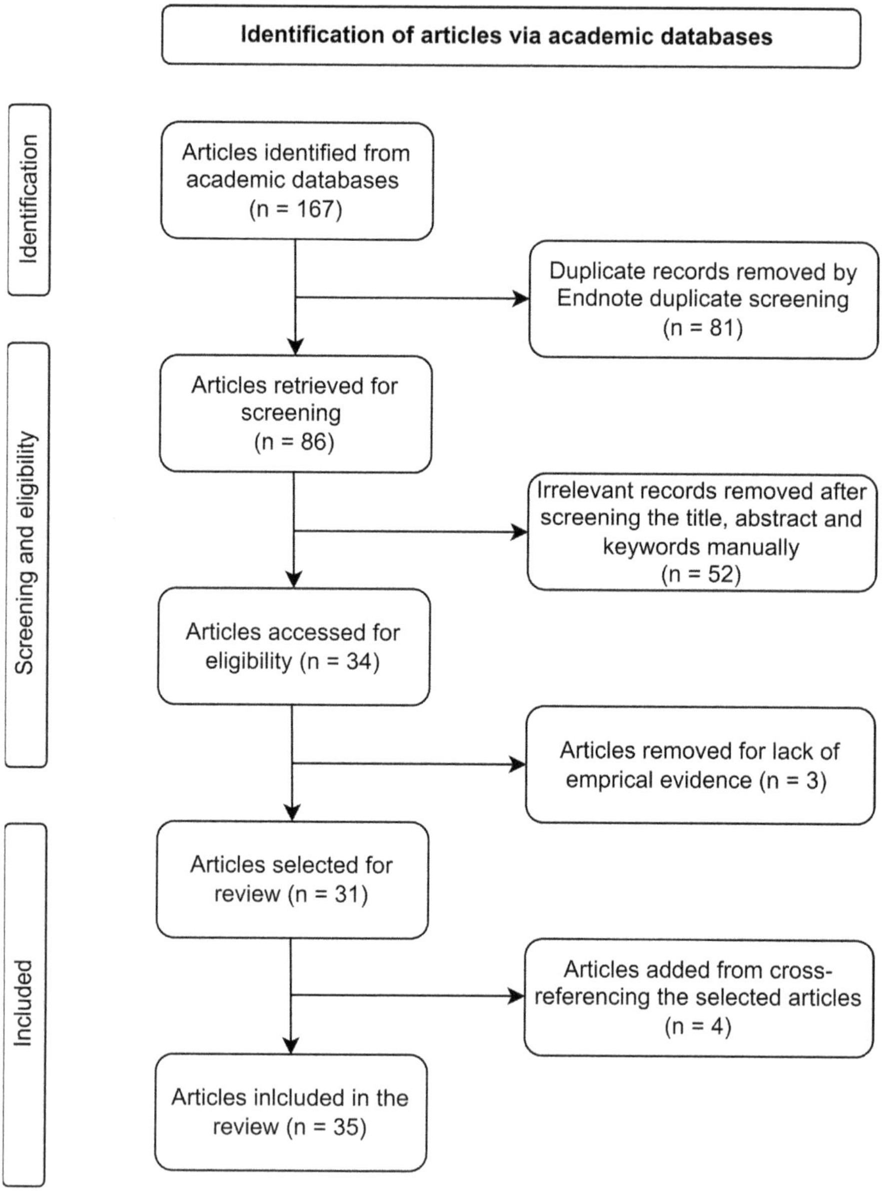

Figure 2.2 Identification of Articles via Databases and Journals

Source: Drafted by the author

were excluded. The screening provided 31 articles meeting the eligibility criteria. Additionally, four articles aligning with the review objectives were retrieved from cross-referencing the filtered articles. Overall, 35 articles were included in the review.

2.6 Role of Information and Communication Technologies in CWQ

A list of 35 articles was retrieved to explore the application of ICTs in quantifying CW generation across the project life cycle stages. The selected articles are grouped into five categories based on the type of ICTs incorporated into the CWQ process. The categories include (i) BIM-integrated CWQ frameworks, (ii) AI-based CW prediction and classification models, (iii) On-site CW segregation and sorting using robots, (iv) Bigdata-enabled WGR benchmarking, and (v) GIS-image integrated CW models (Table 2.4).

2.6.1 BIM-Integrated CWQ Frameworks

BIM, one of the leading data-rich and object-oriented collaborative technologies, is incorporated into the CWQ process. The application of BIM can be seen in three ways: (i) as an information bank for acquiring design, construction, and material information, (ii) as a visualisation tool, and (iii) for its computational API features.

2.6.1.1 As an Information Bank

CWQ is considered challenging due to the complex nature of projects, which requires a detailed understanding of the physical and functional properties of materials [96]. Witnessing the data richness in BIM models, researchers used BIM-based MTO to obtain accurate and real-time dimensional and functional data of buildings [56]. The multi-dimensional capability of BIM represents the building information across the life-cycle stages, ranging from 3D modelling to scheduling (4D), cost planning (5D), sustainability (6D), and facilities management (7D). Cheng and Ma [29] were the first researchers to incorporate 3D BIM to estimate building demolition and renovation waste quantities, pick-up truck requirements, and disposal costs. Later, the incorporation of 3D BIM to obtain material details required for waste estimation can be seen in many studies [61, 62, 64, 97–101]. In addition, a few studies have incorporated higher dimensions of BIM to quantify waste. For instance, a 4D-BIM framework was developed by Guerra et al. [102] to estimate concrete and drywall waste quantities and schedule for its reuse and recycling during building construction. Besides, the 7D BIM data was incorporated into the quantification process to estimate waste volume, pick-up time, cost, and reuse and recycling capabilities [103].

2.6.1.2 As a Visualisation Tool

Earlier studies highlighted the impact of design changes and design errors causing reworks and waste generation during construction, stating that design factors are responsible for one-third of total waste generation [20, 126]. Besides, literature [58, 127] proves that the visualisation and design validation features of BIM facilitate architects and designers to identify clashes and develop error-free designs in the virtual environment, thereby avoiding reworks and wastage on-site. It is proved that 4.3%–15.2% of waste is prevented by integrating the BIM-enabled visualisation features to detect clashes in the design phase [34].

Table 2.4 Existing CWQ Models

Author and year	Location	*Type of activities*				*Units of measure*		*Waste classification type*						*Types of CWQ models*					
		CW	*Re/RfW*	*DW*	*None*	*Weight and volume*	*Impacts*	*A*	*B*	*C*	*D*	*E*	*F*	*1*	*2*	*3*	*4*	*5*	*6*
[29]	HK			✓		✓		✓						✓					
[104]	HK	✓	✓	✓		✓				✓							✓		
[105]	Korea	✓				✓							✓		✓				
[106]	HK	✓				✓							✓		✓		✓		
[34]	South Korea	✓				✓							✓	✓					
[107]	Korea			✓		✓		✓						✓					
[108]	Austria			✓		✓		✓					✓					✓	
[109]	China	✓		✓		✓		✓							✓				
[100]	China			✓			✓	✓						✓					
[62]	UK	✓				✓		✓						✓				✓	
[110]	USA	✓				✓		✓*						✓					
[111]	HK				✓	✓		✓*							✓	✓			
[101]	China	✓					✓	✓						✓					
[19]	UK			✓		✓			✓						✓				
[102]	USA	✓				✓		✓*						✓					
[112]	HK				✓	✓		✓							✓	✓			
[113]	China				✓	✓		✓*							✓				
[114]	Bahrain				✓	✓							✓		✓				
[115]	Australia				✓	✓		✓							✓				
[30]	New Zealand	✓				✓		✓											✓
[116]	China	✓				✓					✓				✓				
[61]	Bangladesh	✓				✓		✓[1]						✓					
[117]	China				✓	✓							✓		✓				
[97]	Spain	✓				✓						✓		✓					

(*Continued*)

Table 2.4 (Continued)

		Type of activities				*Units of measure*		*Waste classification type*						*Types of CWQ models*					
Author and year	*Location*	*CW*	*Re/RfW*	*DW*	*None*	*Weight and volume*	*Impacts*	*A*	*B*	*C*	*D*	*E*	*F*	*1*	*2*	*3*	*4*	*5*	*6*
[99]	China	✓				✓		✓						✓					
[64]	China			✓		✓			✓					✓				✓	
[118]	HK				✓	✓				✓							✓		
[119]	China				✓	✓		✓							✓	✓			
[120]	HK				✓	✓		✓							✓				
[121]	China				✓	✓		✓							✓				
[122]	HK				✓	✓		✓							✓				
[123]	Korea				✓	✓		✓[2]							✓				
[98]	Spain	✓				✓						✓		✓					
[124]	4 cities (1 in Canada and 3 in the USA)				✓	✓							✓					✓[3]	
[125]	Canada				✓	✓							✓					✓[3]	

Legends:

A – Material classification, B – Reusability levels, C – Reactive waste nature (Inert, non-inert and hazardous), D – Chemical properties (inorganic, organic, metallic, and non-metallic), E – EWC, and F – No material classification.

1 – BIM-integrated CWQ frameworks; 2 – AI-based CW prediction and classification models; 3 – On-site CW estimation using robots; 4 – Bigdata-enabled WGR benchmarking; 5 – GIS-image integrated CW models; and 6 – Other waste-data collection and reporting platforms.

Source: Drafted by the author

Notes:
[1] Only specific material(s) are selected for quantification;
[2] Korean classification system; and
[3] GIS tools integrated with satellite image analysis.

2.6.1.3 For Its Computational API Features

In addition to the internal BIM applications, the functionality of BIM authoring tools is expanded to communicate with external applications through APIs and plugins. The model developed by Cheng and Ma [29] was the first research to create a plugin using BIM-API to measure demolition and renovation waste. Also, Akinade and Oyedele [62] developed a plugin to integrate the AI-based waste prediction model with the BIM platform to facilitate total waste estimation and design simulation in the design stages for new buildings. The API benefits the users by automatically extracting the design and material information, estimating waste, and inputting waste information into the model. This API enhances the dynamics between BIM and CWQ models.

2.6.2 AI-Based CW Prediction and Classification Models

AI modelling provides excellent value to various application domains, including the built environment [128–130]. Similarly, various AI models that support CW quantification and classification processes are applied across the project life cycle. The AI models are widely used to generalise patterns from large datasets with less human intervention and high accuracy. The current studies use AI models to analyse and interpret patterns from the building and regional CW datasets. The main features of AI employed in CWQ models include (i) predictive analytics and (ii) image analysis and computer vision.

2.6.2.1 Predictive Analytics

Depending on the project functionalities and the volume of existing data records, AI models are selected to predict/forecast waste generation levels. The two main pillars supporting AI modelling are input data and training algorithm(s). For instance, support vector machines (SVMs) were used to demonstrate the relationship between predictor variables (influencing factors of waste generation) and WGRs of the small data set [116]. The data from 206 ongoing construction projects were used to develop a prediction model. SVM, a supervised ML algorithm, is used for classification, regression, and outlier detection in given datasets [131]. Meanwhile, artificial neural networks (ANNs), a DL model, are used to generalise patterns from large datasets. For instance, Lu and Peng [132] used an ANN to predict CW generation at early design stages. The waste patterns were interpreted from 37K waste disposal records of multifamily residential buildings in Hong Kong. The results of the study emphasise that WGRs are highly influenced by geographical location, project duration, project cost and public-private nature.

However, accuracy, time, and estimation complexities are significant shortcomings of single AI models [105]. Hence, a few studies [19, 133] developed hybrid AI models combining two algorithms to quantify CW. The application of hybrid models allowed the researchers to overcome the limits of individual models and benefit from combining their strengths. For instance, Akinade and Oyedele [62] modelled a hybrid model, a combination of neural networks and fuzzy inference systems, to develop an ideal neural network structure within a limited time, which otherwise would have been a significant shortcoming in the homogenous model. Furthermore, a hybrid model combining ANNs with ant colony optimisation was proposed to estimate waste in the early construction stages of residential buildings [105].

2.6.2.2 *Image Analysis and Computer Vision*

Besides, AI algorithms are integrated into waste analysis to discover patterns from digital waste images, thus facilitating on-site waste picking, classification, and sorting [113, 115, 120–123]. Using AI models to recognise images enhances the classification and quantification of CW generated. The image data required for model training is mainly captured from waste skips placed on-site [115], waste dumps [123], truckload images [122, 134], and Google search images [121]. Depending on the type of image selected, the spatial features, visual aspects, and boundaries of waste materials are extracted to design the classification models. For instance, a research study conducted by Dong et al. [120] developed a DL-integrated boundary transformer framework based on data collected from waste skips. Thus, integrating the semantic segmentation and object detection features of computer vision, this automated framework facilitates waste recognition in a cluttered state and supplements a fine granularity of material information within the given boundaries. Therefore, this reduces human intervention, enhances waste reporting, and allows waste handlers to assign waste codes based on material types to undergo safe disposal.

Based on the analysis, DL models like ANN and convolution neural networks have proven more appropriate for working with unstructured and fuzzy waste data. However, the availability and quality of waste records are significant constraints to utilising AI's potential in the CWQ process. Accounting for the data limitation, the study by [123] employed data augmentation techniques to create new data points from the existing data to amplify the dataset. Thus, this amplification in this study [123] reduced the overfitting of models and increased generalisation, thereby increasing the applications and accuracy of DL models. However, to utilise the potential of AI models, it is significant to ensure data quality, accuracy, and consistency.

2.6.3 *On-Site CW Segregation and Sorting Using Robots*

Since CW comprises diverse materials, each has a unique physical and chemical composition that influences the overall resource value of waste. So, segregating materials at the source eliminates contamination and increases the recycling and recovery potentials of CW [135]. Nevertheless, handling these waste materials is time-consuming, cost-ineffective, and requires manual work [136]. Besides, when handled manually, hazardous substances within the CW pose significant threats to health and safety. To overcome these issues, autonomous robots are designed to pick up waste and sort it according to the material type. Research studies [111, 112, 119] integrated computer vision algorithms to design robots that support waste picking and on-site sorting. Several robotic models [133, 137] are designed to facilitate CW sorting in waste treatment facilities.

Despite the benefits, robotic models face challenges in recognising the waste material type, thus causing quality issues. Hence, to overcome these gaps, a recent study by [138] proposed a human-robot collaborative model by integrating the functionalities of AR with Robots. This integration takes advantage of robots doing repetitive sorting and humans using their cognitive skills to enhance the CW sorting quality. The AR system within the model allows humans to monitor and modify the information required to enhance the quality and efficiency of robot-enabled automatic CW sorting. The study benefited from using robotic functionalities to reduce manual work on repetitive tasks and used human cognitive skills to enhance waste sorting. Using robots on-site for waste picking and sorting is expected to minimise mixed-waste quantities, thereby increasing the waste reuse and recycling potentials. Furthermore, automated sorting is advised to help the built environment overcome the health and safety accidents associated with manual waste handling.

2.6.4 Bigdata-Enabled WGR Benchmarking

Big data, a data-driven decision-making approach, is adopted to measure and benchmark CW performance. For instance, the big data was used to benchmark CWM performance using consistent WGRs for construction, renovation, and demolition works [104]. This study classified waste as inert and non-inert and used a statistical approach to analyse patterns of those waste types. Likewise, another study by Lee, Kim [105] proposed an analytical model linked with AI and big data to discover relationships between project characteristics and automate the quantification process. Apart from waste estimation, the role of big data in waste classification is also demonstrated in the literature. For instance, a big data model was developed by Yuan, Lu [118] using 4.27 million truckload data to estimate the material composition of waste. The model development was facilitated by combining the statistical and probability approaches with big-data analytics.

2.6.5 GIS-Integrated CW Models

GIS is a spatial system that deals with data capture, storage, integration, and spatial and location information analysis. GIS models are developed by [64, 108, 139, 140] to acquire data required for quantification. For instance, Wu, Wang [139] used GIS images to estimate demolition waste generation at the regional level. The building usage type, type of structural frame, Gross Floor Area (GFA), and service life of each building were collected using ArcGIS to measure the expected waste generation levels. The same application can be seen in measuring demolition waste during the urban renewal process in China [141]. Further, the GIS databases and remote sensing techniques were utilised to quantify CW of existing building stock [108]. The study shows that statistical analysis underestimates the gross volume of demolished buildings by 1.1 million m^3 per annum across a specific geographical area, which causes inaccuracy in the waste outputs. It has also been proved that errors occurring from other building records can be mitigated by incorporating GIS images into CWQ [108]. In addition to measuring building layouts, the site locations and spatial features of waste recycling plants and landfill sites were also analysed using GIS maps to facilitate efficient waste treatment and disposal [64]. In another study [125], the GIS tools were integrated with satellite images to measure the CW disposal rates at the regional level. The images of a specific landfill site were captured to identify the changes in the settlement area during 2014–2022. This approach also allowed the authors to identify the influence of socioeconomic factors on CW disposal rates. The satellite images were also used in other studies like [124] to estimate the CW generation rates.

2.7 Limitations of Existing CWQ Frameworks

Although several CWQ models are presented in the literature, it is essential to understand the critical limitations restraining its applications. The five significant limitations include (i) ununified and unstandardised data structure leading to interoperability issues, (ii) complex relationships between project characteristics and WGRs are not defined, (iii) bidirectional connectivity between BIM and CW platforms are deficient, (iv) inadequate assessment of CW impacts, (v) lack of information granularity, and (vi) whole life-cycle assessments of CW are lacking, (vii) lack of focus on refurbishment and renovation activities, and (viii) limited attention given to infrastructure and civil engineering works.

2.7.1 Ununified and Unstandardised Data Structure Impeding Interoperability

The role of ICTs like BIM and AI is increasingly implemented in CWQ for its benefits across data collection, predictive analysis, modelling, simulation, and visualisation. Despite its advantages, the continuous evolution of computational systems has led to the deployment of heterogeneous data and technologies, which will keep growing in the future. For instance, accounting BIM as a source for building information, the outputs of several databases including site-waste reports, historical and regional waste databases, GIS, and AI model outputs are integrated to different extents to facilitate CWQ. Thus, acquiring large amounts of heterogeneous data from multiple sources to assess waste holds different data structures, significantly impacting integration and information exchange issues [142]. The ununified data structure compromises data accuracy, information consistency, and timely decision-making. Hence, a unified data architecture contextualising diversified data from multiple sources would enhance interoperability and facilitate more robust and granular analysis. This unification further allows AI-driven automation and improves the dynamics between multiple data platforms. Besides, the unified data structure enables the integration of CW analysis with actual project workflows, including time, cost, and other supply chain indicators that aid in robust and collaborative decision-making.

2.7.2 Complex Relationships Between Project Factors and WGRs Are Unspecified

The productivity of construction projects is progressively improved by adopting automated and flexible techniques, which reduces human intervention across the project life-cycle stages [143]. Similarly, different levels of digital applications are deployed to facilitate the quantification and classification of CW. The CWQ process is primarily modelled with two significant factors: the input (building characteristics) and output (WGRs) variables. The building characteristics used in the existing CWQ models are GFA, building usage type, and structural frame. However, in real cases, WGR is influenced by various other project factors, including architectural form, procurement type, stages of construction, installation methods, and temporal aspects [20, 31, 32, 144]. For instance, in conventional construction, the super-structure stage generates a large volume of steel reinforcement, brick, and block waste due to offcuts. These waste types are prevented in prefabricated structures, but the volume of plastic packing waste used to transport these components increases [35], further becoming a concern on increased hazardous waste generation. Despite the similar building usage and structural frames, this hypothesis shows that the WGR of brick, block, and steel varies between conventional and prefabricated constructions. Hence, adopting the same WGRs in both cases could lead to inaccurate results. Therefore, a comprehensive analysis of CW and its inherent relationship with multiple building design and construction features, in addition to the basic characteristics, is paramount to enhancing the granularity and appropriateness of the results.

2.7.3 Bidirectional Connectivity Between BIM and CW Platforms Is Deficient

Another emerging research direction is the enhancement of the interfaces between the CWQ models and BIM platforms. BIM is widely used in the built environment for its numerous benefits. This includes its use as a data repository for holding detailed information on the physical and functional properties of building elements, its promotion of collaboration through a common data environment, and its enhancement of data sharing across different parties [56, 58]. The multi-dimensional property of BIM holds piles of information required for improving the overall CWQ process. However, most existing BIM-incorporated quantification solutions are stand-alone and unidirectional, lacking integration with the BIM platforms, except for a few approaches [29, 62]. Interoperability problems could result from exchanging data between various systems, significantly hindering the digital construction age. This static approach also compromises options to visualise the real-time CW levels caused by design and construction factors. So, a promising future research direction could involve exploring the bidirectional connectivity between the CWQ models and BIM platforms through API features, thus enhancing visualisation and real-time evaluation. In addition, leveraging the CW workflows with BIM could support the project team in monitoring and allocating resources on time and within budget.

2.7.4 Inadequate Assessment of CW Impacts

Each CWQ model is unique in the way they measure and report waste. WGR is widely used in existing research studies to quantify and benchmark CW performance. Based on the WGRs of building materials, it was suggested in the reviewed studies that inert waste materials contribute to a large volume of CW, and the quantity of hazardous waste is negligible [116, 145]. According to the articles evaluated, the primary waste-contributing materials in terms of volume/weight are concrete, bricks/blocks, ceramics, and timber [10–12, 108, 146, 147]. However, a few studies [11, 100, 101] observed that emission levels of other waste materials like metals, gypsum, and asphalt are higher than the 'major waste-contributing' materials. For instance, Xu, Shi [101] estimated that asphalt contributed to one-fourth of the carbon emissions, although the material was the least ranked in waste quantities. The recycling of aluminium waste contributes to 45% of carbon emissions despite its weight being 1% of total waste generation. This scenario is vice versa in masonry waste, which contributes to 90% of the total waste volume, but its emission levels are less than 1% of total emissions [100]. Therefore, reducing emission levels (embodied carbon levels) based on the conclusions reached using volume- and weight-based waste results will have limited impacts on achieving net zero carbon targets.

In addition, economic aspects of waste, as evidenced by the higher tax pricing [49] for hazardous waste (£103.70/tonne) compared to inert waste (£3.30/tonne), suggest that the former should also be prioritised within the waste flows. Thus, future studies should exploit incorporating the impact values of waste materials during quantification. Besides, impact-based analysis will seamlessly integrate the CWQ models with the design, construction, carbon, and cost workflows.

2.7.5 Lack of Information Granularity

Classifying waste based on its material types is essential to enhance the quantification and optimisation accuracy. As a result, most studies adopted a detailed waste classification approach (Table 2.3). However, the terms used to describe the composition/distribution of waste materials have been inconsistent, compromising automation and seamless information exchange. For instance, a few studies [146, 148] used the terms 'masonry/masonry debris/masonry blocks' to report the waste generation of 'bricks/blocks'. Likewise, 'plasterboard' is called 'drywall/gypsum wallboard/wall panels' [110, 149] based on geographical location. Although the properties of these materials are similar, there is a less semantic definition for these terms to make informed decisions. With the growing usage of ICT techniques, research studies [150] determine that reliable waste-data representation is essential for lowering errors brought on by inconsistent definitions. A few studies, including [116], maintained a uniform structure to integrate data collected from various projects for quantification. Nevertheless, the representation of defined vocabularies to automate the overall system is lacking.

Besides, the literature [97, 98] presents standard waste code systems as a viable solution to enhance the semantic characterisation of waste materials. For instance, in the studies conducted by [97, 98], the Base de Costes de la Construcción de Andalucía (BCCA)-enabled coding system was used to map data between the BIM library and the selected CW database. However, this system is specific to the study and lacks applications in a broader context. Hence, initiatives incorporating standardised material and waste codes would ease information exchange, mitigate communication issues, and ensure automation across the CWQ process. These functions further enable a dynamic integration with BIM platforms.

2.7.6 Whole Life-Cycle Assessments of CW Are Lacking

LCA is a methodology used to measure the environmental impacts of materials from their extraction to disposal, called 'cradle to grave' analysis [151]. It is acknowledged as an established method for making informed decisions on selecting an appropriate WM route based on its environmental impacts [100]. However, the primary focus of the studies was on specific project execution stages like construction, operation, maintenance, and demolition (Table 2.3) while ignoring the other life cycle stages. It is vital to note that the entire life cycle of a material includes product manufacturing stages and stages after its EoL in addition to its application and use across various project life-cycle stages. Thus, sustainable materials are the ones that have less impact not only on the specific stage(s) but also across the whole material life cycle. The importance of encompassing the whole life cycle in CW assessment is also highlighted by [15]. The report states that the scope of CWM should be across the whole life cycle of a material. This, therefore, suggests that the entire life cycle of the material should be accounted for while making CW-informed decisions.

2.7.7 Lack of Focus on Refurbishment/Renovation Activities

Existing CWQ studies have focused on new constructions and demolition activities, while refurbishments and renovations of existing structures have received limited attention. This is despite the WGR of refurbishment/renovation and demolition works being three times higher than those of new construction [152]. Furthermore, national laws on net-zero carbon targets appear to have driven the reuse of existing structures in some countries for their potential benefits across energy, cost, social, and health values. For instance, 70% of the UK's non-residential building stock is already built and will require refurbishing one million homes yearly for the next three decades to meet the 2050 net-zero target [153]. Waste generation during the in-use stage increases as refurbishment and renovation activities expand, thus increasing the need for a comprehensive CWQ in those stages. This suggests the demand for a shift in focus; waste from renovation and refurbishment should be considered a crucial component of WQ practices.

Besides, the unavailability of waste data to measure the EoL performance of a building acts as a significant barrier to their detailed assessment [28, 154, 155]. Unlike new buildings, existing buildings are not designed using BIM. Most existing buildings neither hold digital information nor incorporate EoL considerations during their design and construction, so there is a scant level of data available to estimate the reuse and recovery potential of CW precisely. The need to propose a retrospective BIM model of existing buildings should be prioritised in future studies to integrate the EoL considerations in quantifying CW and informing optimal choices.

2.7.8 Limited Attention on Infrastructure and Civil Engineering Works

Unlike buildings, the studies focusing on WQ from infrastructure works are deficient. Among the evaluated studies, only one paper [156] focused on measuring waste from infrastructure (railway) projects in Spain. Given the wide variety of materials used in buildings in the real world, a thorough analysis of waste from buildings is essential [108, 157]. However, it has been suggested that infrastructure projects also produce significant waste because of their size and complexity, which must be prioritised for analysis [158]. In addition, waste-generating factors, including design features, construction techniques, and other environmental and social factors, impact the infrastructure. So, adequate planning in the early project stages is significant for controlling the waste generation in the infrastructure works [158]. This indicates a potential field for future scientific research to concentrate on measuring waste from infrastructure projects.

2.8 Summary

The definition of waste, its types, impacts on the triple bottom line of sustainability, and factors influencing waste generation across the life-cycle stages are discussed in this chapter. The highlights from the chapter are as follows.

CW generation and types:

- The material waste generated from building and infrastructure projects is distributed across the supply chain.
- There are eight different types of factors used for the classification of CW. It includes the reactive nature of waste materials, material types, EWC codes, life cycle stages and origin and characteristics of waste at each life-cycle stage, activities associated with construction, maintenance, and demolition of buildings, the origin of material sources, and WM routes. The type of classification approach depends on the purpose of the study.
- The factors within the design phase contribute to one-third of the total waste generation in buildings.
- The CWM is widely supported by advancements in ICTs and changes in resource consumption, design, construction, deconstruction, and reuse. Legal policies and regulatory frameworks further support these changes.

CWQ:

- Quantification is the process of assessing the quantity of waste generated. The scope of quantification associated with the research is project-level CWQ quantification for buildings.
- A list of 35 CWQ models that met the SLR eligibility criteria was retrieved for review.
- The retrieved CWQ models are grouped into five categories based on the type of ICTs used. The categories include (i) BIM-integrated CWQ frameworks,

(ii) AI-based CW prediction and classification models, (iii) On-site CW segregation and sorting using robots, (iv) WGR benchmarking using big data, and (v) GIS-image integrated CW models.

- The benefits identified by integrating ICTs in CWQ are that it minimises the level of human intervention in the quantification process and increases the accuracy of quantification results.
- The limitations of existing CWQ models include ununified and unstandardised data structure leading to interoperability issues, complex relationships between project characteristics and WGRs are not defined, bidirectional connectivity between BIM and CW platforms are deficient, inadequate assessment of CW impacts, lack of information granularity, whole life-cycle assessments of CW are lacking, lack of focus on refurbishment/renovation activities, and limited study on infrastructure and civil engineering works.
- The outcomes of the book chapter will assist researchers in identifying the prevailing gaps and enhancing the CWQ workflows to meet digital and information demands in the built environment.

References

1. EUROSTAT, *Generation of waste by economic activity*, 2022. https://ec.europa.eu/eurostat/databrowser/view/ten00106/default/table?lang=en. Last access: 04 January 2023.
2. Defra, *UK statistics on waste*, 2022. https://www.gov.uk/government/statistics/uk-waste-data/uk-statistics-on-waste. Last access: 04 January 2023.
3. S. Sivashanmugam, R. Sergio, R. Farzad and D. Nashwan, Maximising the construction waste reduction potential – Barriers and catalysts. *Proceedings of the Institution of Civil Engineers – Civil Engineering*, 2022. 176(6): pp. 1–9. https://doi.org/10.1680/jcien.22.00163
4. M.B. Purushothaman and J. Seadon, System-wide construction waste and their connectivity to construction phases, impacting 5M factors and effects: A systematic review. *Smart and Sustainable Built Environment*, 2024. 13(2): pp. 354–369. https://doi.org/10.1108/SASBE-12-2022-0269
5. EC Directive, *Directive 2008/98/EC of the European Parliament and of the Council of 19 November 2008 on waste and repealing certain directives*, 2008. https://eur-lex.europa.eu/legal-content/EN/TXT/?uri=celex%3A32008L0098. Last access: 28 June 2023.
6. US EPA, *Sustainable management of construction and demolition materials*, 2024. https://www.epa.gov/smm/sustainable-management-construction-and-demolition-materials#:~:text=Construction%20and%20Demolition%20(C%26D)%20materials,buildings%2C%20roads%2C%20and%20bridges. Last access: 20 October 2024.
7. S. Sivashanmugam, S. Rodriguez Trejo and F. P. Rahimian, BIM-integrated semantic framework for construction waste quantification and optimisation. *Automation in Construction*, 2024. 168: p. 105842. https://doi.org/10.1016/j.autcon.2024.105842
8. S. Sivashanmugam, S. Rodriguez, F.P. Rahimian, F. Elghaish and N. Dawood, Enhancing information standards for automated construction waste quantification and classification. *Automation in Construction*, 2023. 152: p. 104898. https://doi.org/10.1016/j.autcon.2023.104898
9. F. Kleemann, J. Lederer, H. Rechberger and J. Fellner, GIS-based analysis of Vienna's material stock in buildings. *Journal of Industrial Ecology*, 2017. 21(2): pp. 368–380. https://doi.org/10.1111/jiec.12446

10. J. Li, Z. Ding, X. Mi and J. Wang, A model for estimating construction waste generation index for building project in China. *Resources, Conservation and Recycling*, 2013. 74: pp. 20–26. https://doi.org/10.1016/j.resconrec.2013.02.015
11. P. Sun, N. Zhang, J. Zuo, R. Mao, X. Gao and H. Duan, Characterizing the generation and flows of building interior decoration and renovation waste: A case study in Shenzhen City. *Journal of Cleaner Production*, 2020. 260: p. 121077. https://doi.org/10.1016/j.jclepro.2020.121077
12. H. Wu, H. Duan, L. Zheng, J. Wang, Y. Niu and G. Zhang, Demolition waste generation and recycling potentials in a rapidly developing flagship megacity of South China: Prospective scenarios and implications. *Construction and Building Materials*, 2016. 113: pp. 1007–1016. https://doi.org/10.1016/j.conbuildmat.2016.03.130
13. European Commission (EC), *Guidance on classification of waste according to EWC-Stat categories*, 2010. https://ec.europa.eu/eurostat/documents/342366/351806/Guidance-on-EWCStat-categories−2010.pdf/0e7cd3fc-c05c-47a7–818f-1c2421e55604. Last access: 04 January 2023.
14. Environment Agency, *Guidance on the classification and assessment of waste* (1st Edition v1.2.GB) – Technical Guidance WM3, 2021. https://assets.publishing.service.gov.uk/government/uploads/system/uploads/attachment_data/file/1021051/Waste_classification_technical_guidance_WM3.pdf. Last access: 02 June 2023.
15. CLC, *Zero avoidable waste in construction*, 2020. https://www.constructionleadershipcouncil.co.uk/wp-content/uploads/2016/05/ZAW-Report-Final-Draft-25-February−2020.pdf. Last access: 15 May 2023.
16. M.F. Hasmori, A.F.M. Zin, S. Nagapan, R. Deraman, N. Abas, R. Yunus and M. Klufallah, The on-site waste minimization practices for construction waste, in *IOP conference series: Materials science and engineering*. 2020. IOP Publishing. https://doi.org/10.1088/1757–899X/713/1/012038
17. C.S. Poon, A.T.W. Yu, S.W. Wong and E. Cheung, Management of construction waste in public housing projects in Hong Kong. *Construction Management & Economics*, 2004. 22(7): pp. 675–689. https://doi.org/10.1080/0144619042000213292
18. Q. Wang, L. Chen, R. Hu, Z. Ren, Y. He, D. Liu and Z. Zhou, An empirical study on waste generation rates at different stages of construction projects in China. *Waste Management & Research*, 2020. 38(4): pp. 433–443. https://doi.org/10.1177/0734242x19886635
19. L.A. Akanbi, A.O. Oyedele, L.O. Oyedele and R.O. Salami, Deep learning model for demolition waste prediction in a circular economy. *Journal of Cleaner Production*, 2020. 274: p. 122843. https://doi.org/10.1016/j.jclepro.2020.122843
20. B. Bossink and H. Brouwers, Construction waste: Quantification and source evaluation. *Journal of Construction Engineering and Management*, 1996. 122(1): pp. 55–60. https://doi.org/10.1061/(ASCE)0733–9364(1996)122:1(55)
21. M. Osmani, Design waste mapping: A project life cycle approach, in *Proceedings of the institution of civil engineers-waste and resource management*, 2013. ICE Publishing. https://doi.org/10.1680/warm.13.00013
22. M. Osmani, Construction waste minimization in the UK: Current pressures for change and approaches. *Procedia-Social and Behavioral Sciences*, 2012. 40: pp. 37–40. https://doi.org/10.1016/j.sbspro.2012.03.158
23. C.T. Formoso, L. Soibelman, C. De Cesare and E.L. Isatto, Material waste in building industry: Main causes and prevention. *Journal of Construction Engineering and Management*, 2002. 128(4): pp. 316–325. https://doi.org/10.1061/(ASCE)0733–9364(2002)128:4(316)
24. D. Gann, A. Salter and J. Whyte, Design quality indicator as a tool for thinking. *Building Research & Information*, 2003. 31(5): pp. 318–333. https://doi.org/10.1080/0961321032000107564
25. L.L. Ekanayake and G. Ofori, Building waste assessment score: Design-based tool. *Building and Environment*, 2004. 39(7): pp. 851–861. https://doi.org/10.1016/j.buildenv.2004.01.007

26. S.O. Ajayi and L.O. Oyedele, Critical design factors for minimising waste in construction projects: A structural equation modelling approach. *Resources, Conservation and Recycling*, 2018. 137: pp. 302–313. https://doi.org/10.1016/j.resconrec.2018.06.005
27. H. Yuan, Critical management measures contributing to construction waste management: Evidence from construction projects in China. *Project Management Journal*, 2013. 44(4): pp. 101–112. https://doi.org/10.1002/pmj.21349
28. O.O. Akinade, L.O. Oyedele, S.O. Ajayi, M. Bilal, H.A. Alaka, H.A. Owolabi and O.O. Arawomo, Designing out construction waste using BIM technology: Stakeholders' expectations for industry deployment. *Journal of Cleaner Production*, 2018. 180: pp. 375–385. https://doi.org/10.1016/j.jclepro.2018.01.022
29. J.C. Cheng and L.Y. Ma, A BIM-based system for demolition and renovation waste estimation and planning. *Waste Management*, 2013. 33(6): pp. 1539–1551. https://doi.org/10.1016/j.wasman.2013.01.001
30. N. Domingo and T. Batty, Construction waste modelling for residential construction projects in New Zealand to enhance design outcomes. *Waste Management*, 2021. 120: pp. 484–493. https://doi.org/10.1016/j.wasman.2020.10.010
31. S. Nagapan, I.A. Rahman, A. Asmi, A.H. Memon and I. Latif, Issues on construction waste: The need for sustainable waste management, in *2012 IEEE colloquium on humanities, science and engineering (CHUSER)*, 2012. IEEE. https://doi.org/10.1109/CHUSER.2012.6504333
32. M. Osmani, J. Glass and A.D. Price, Architects' perspectives on construction waste reduction by design. *Waste Management*, 2008. 28(7): pp. 1147–1158. https://doi.org/10.1016/j.wasman.2007.05.011
33. C.-S. Poon, A.T. Yu and L. Jaillon, Reducing building waste at construction sites in Hong Kong. *Construction Management and Economics*, 2004. 22(5): pp. 461–470. https://doi.org/10.1080/0144619042000202816
34. J. Won, J.C. Cheng and G. Lee, Quantification of construction waste prevented by BIM-based design validation: Case studies in South Korea. *Waste Management*, 2016. 49: pp. 170–180.
35. N.G. Pericot, P.V. Sáez, M.D.R. Merino and O.L. Carrasco, Production patterns of packaging waste categories generated at typical Mediterranean residential building worksites. *Waste Management*, 2014. 34(11): pp. 1932–1938. https://doi.org/10.1016/j.wasman.2014.06.020
36. V. Swarnakar and M. Khalfan, Circular economy in construction and demolition waste management: An in-depth review and future perspectives in the construction sector. *Smart and Sustainable Built Environment*, 2024 (ahead-of-print). https://doi.org/10.1108/SASBE-02-2024-0056
37. Z. Liu, M. Osmani, P. Demian and A. Baldwin, A BIM-aided construction waste minimisation framework. *Automation in Construction*, 2015. 59: pp. 1–23. https://doi.org/10.1016/j.autcon.2015.07.020
38. Office for National Statistics, *Building materials and components statistics: October 2022*, 2022. https://www.ons.gov.uk/economy/environmentalaccounts/datasets/ukenvironmentalaccountsmaterialflowsaccountunitedkingdom/current. Last access: 07 December 2024.
39. CLC and The Green Construction Board, *The routemap for zero avoidable waste in construction*, 2021. https://www.constructionleadershipcouncil.co.uk/wp-content/uploads/2021/07/ZAW-Interactive-Routemap-FINAL.pdf. Last access: 07 December 2024.
40. A. Padilla-Rivera, B. Amor and P. Blanchet, Evaluating the link between low carbon reductions strategies and its performance in the context of climate change: A carbon footprint of a wood-frame residential building in Quebec, Canada. *Sustainability*, 2018. 10(8): p. 2715. https://doi.org/10.3390/su10082715
41. W.K. Chong and C. Hermreck, Understanding transportation energy and technical metabolism of construction waste recycling. *Resources, Conservation and Recycling*, 2010. 54(9): pp. 579–590. https://doi.org/10.1016/j.resconrec.2009.10.015

42. A. Coelho and J. De Brito, Influence of construction and demolition waste management on the environmental impact of buildings. *Waste Management*, 2012. 32(3): pp. 532–541. https://doi.org/10.1016/j.wasman.2011.11.011
43. T. Ann, C.S. Poon, A. Wong, R. Yip and L. Jaillon, Impact of construction waste disposal charging scheme on work practices at construction sites in Hong Kong. *Waste Management*, 2013. 33(1): pp. 138–146. https://doi.org/10.1016/j.wasman.2012.09.023
44. O.O. Fadiya, P. Georgakis and E. Chinyio, Quantitative analysis of the sources of construction waste. *Journal of Construction Engineering*, 2014. 2014: p. 9. https://doi.org/10.1155/2014/651060
45. M. Marzouk and S. Azab, Environmental and economic impact assessment of construction and demolition waste disposal using system dynamics. *Resources, Conservation and Recycling*, 2014. 82: pp. 41–49. https://doi.org/10.1016/j.resconrec.2013.10.015
46. S. Tafesse, Y.E. Girma and E. Dessalegn, Analysis of the socio-economic and environmental impacts of construction waste and management practices. *Heliyon*, 2022. 8(3): p. e09169. https://doi.org/10.1016/j.heliyon.2022.e09169
47. H. Yuan, L. Shen, J.J. Hao and W. Lu, A model for cost–benefit analysis of construction and demolition waste management throughout the waste chain. *Resources, Conservation and Recycling*, 2011. 55(6): pp. 604–612. https://doi.org/10.1016/j.resconrec.2010.06.004
48. BRE, *Construction resources and waste roadmap 2008*, 2008. https://www.bre.co.uk/filelibrary/pdf/rpts/waste/Roadmap_final.pdf. Last access: 06 January 2023.
49. HMRC, *Environmental Taxes Bulletin historical rates*, 2022. https://www.gov.uk/government/statistics/environmental-taxes-bulletin/environmental-taxes-bulletin-historical-rates. Last access: 06 January 2023.
50. K. Lawani, B. Hare and I. Cameron, The true cost of construction material wastage: A case study, in *CIB world building congress 2022*. 2022. https://researchonline.gcu.ac.uk/en/publications/the-true-cost-of-construction-material-wastage-a-case-study#:~:text=The%20true%20cost%20of%20construction%20materials%20wastage%20for%20a%20typical,to%20reflect%20the%20current%20practices
51. S.A. Hosseinijou, S. Mansour and M.A. Shirazi, Social life cycle assessment for material selection: A case study of building materials. *The International Journal of Life Cycle Assessment*, 2014. 19(3): pp. 620–645. https://doi.org/10.1007/s11367-013-0658-1
52. M. Yeheyis, K. Hewage, M.S. Alam, C. Eskicioglu and R. Sadiq, An overview of construction and demolition waste management in Canada: A lifecycle analysis approach to sustainability. *Clean Technologies and Environmental Policy*, 2013. 15(1): pp. 81–91. https://doi.org/10.1007/s10098-012-0481-6
53. H. Yuan, A model for evaluating the social performance of construction waste management. *Waste Management*, 2012. 32(6): pp. 1218–1228. https://doi.org/10.1016/j.wasman.2012.01.028
54. A.P. Dale, P. Chapman and M.L. McDonald, Social impact assessment in Queensland: Why practice lags behind legislative opportunity. *Impact Assessment*, 1997. 15(2): pp. 159–179. https://doi.org/10.1080/07349165.1997.9726130
55. M.U. Hossain, S.T. Ng, P. Antwi-Afari and B. Amor, Circular economy and the construction industry: Existing trends, challenges and prospective framework for sustainable construction. *Renewable and Sustainable Energy Reviews*, 2020. 130: p. 109948. https://doi.org/10.1016/j.rser.2020.109948
56. S. Azhar, Building information modeling (BIM): Trends, benefits, risks, and challenges for the AEC industry. *Leadership and Management in Engineering*, 2011. 11 (3): pp. 241–252. https://doi.org/10.1061/(ASCE)LM.1943–5630.0000127
57. V.W. Tam, C.M. Tam, S. Zeng and W.C. Ng, Towards adoption of prefabrication in construction. *Building and Environment*, 2007. 42(10): pp. 3642–3654. https://doi.org/10.1016/j.buildenv.2006.10.003
58. J. Won and J.C. Cheng, Identifying potential opportunities of building information modeling for construction and demolition waste management and minimization. *Automation in Construction*, 2017. 79: pp. 3–18. https://doi.org/10.1016/j.autcon.2017.02.002

59. H. Yuan, W. Lu and J.J. Hao, The evolution of construction waste sorting on-site. *Renewable and Sustainable Energy Reviews*, 2013. 20: pp. 483–490. https://doi.org/10.1016/j.jclepro.2012.08.016
60. A. Porwal and K.N. Hewage, Building information modeling–based analysis to minimize waste rate of structural reinforcement. *Journal of Construction Engineering and Management*, 2012. 138(8): pp. 943–954. https://doi.org/10.1061/(ASCE)CO.1943–7862.0000508
61. M.T.H. Khondoker, Automated reinforcement trim waste optimization in RC frame structures using building information modeling and mixed-integer linear programming. *Automation in Construction*, 2021. 124: p. 103599. https://doi.org/10.1016/j.autcon.2021.103599
62. O.O. Akinade and L.O. Oyedele, Integrating construction supply chains within a circular economy: An ANFIS-based waste analytics system (A-WAS). *Journal of Cleaner Production*, 2019. 229: pp. 863–873. https://doi.org/10.1016/j.jclepro.2019.04.232
63. Z. Liu, T. Wu, F. Wang, M. Osmani and P. Demian, Blockchain enhanced construction waste information management: A conceptual framework. *Sustainability*, 2022. 14(19): p. 12145. https://doi.org/10.3390/su141912145
64. S. Su, S. Li, J. Ju, Q. Wang and Z. Xu, A building information modeling-based tool for estimating building demolition waste and evaluating its environmental impacts. *Waste Management*, 2021. 134: pp. 159–169. https://doi.org/10.1016/j.wasman.2021.07.025
65. M. Omer, Y.C. Wang, M. Quintana Roma, S. Bedrich, V. Nežerka, J. Ferriz-Papi, J.J. Moros Montanes and I. Diez Ortiz, Development of a 3D Digital model of end-of-service-life buildings for improved demolition waste management through automated demolition waste audit. *Environments*, 2024. 11(7): p. 138. https://doi.org/10.3390/environments11070138
66. A. Talla and S. McIlwaine, Industry 4.0 and the circular economy: Using design-stage digital technology to reduce construction waste. *Smart and Sustainable Built Environment*, 2024. 13(1): pp. 179–198. https://doi.org/10.1108/SASBE-03-2022-0050
67. WRAP, *Designing out waste: A design team guide for buildings*, 2009. https://build360.ie/wp-content/uploads/2023/01/WRAP_DOW_Guide.pdf#page=60.07. Last access: 10 October 2024.
68. L. Jaillon, C.-S. Poon and Y.H. Chiang, Quantifying the waste reduction potential of using prefabrication in building construction in Hong Kong. *Waste Management*, 2009. 29(1): pp. 309–320. https://doi.org/10.1016/j.wasman.2008.02.015
69. W. Lu, W.M. Lee, F. Xue and J. Xu, Revisiting the effects of prefabrication on construction waste minimization: A quantitative study using bigger data. *Resources, Conservation and Recycling*, 2021. 170: p. 105579. https://doi.org/10.1016/j.resconrec.2021.105579
70. J.M. Allwood, J.M. Cullen, M.A. Carruth, D.R. Cooper, M. McBrien, R.L. Milford, M.C. Moynihan and A.C. Patel, *Sustainable materials: With both eyes open*, Vol. 2012, 2012. UIT Cambridge Limited Cambridge, UK. ISBN: 190686005X.
71. R.F. Aziz and S.M. Hafez, Applying lean thinking in construction and performance improvement. *Alexandria Engineering Journal*, 2013. 52(4): pp. 679–695. https://doi.org/10.1016/j.aej.2013.04.008
72. P. Ghisellini, C. Cialani and S. Ulgiati, A review on circular economy: The expected transition to a balanced interplay of environmental and economic systems. *Journal of Cleaner Production*, 2016. 114: pp. 11–32. https://doi.org/10.1016/j.jclepro.2015.09.007
73. UNEP, *Decoupling natural resource use and environmental impacts from economic growth*, 2011. UNEP/Earthprint. https://www.resourcepanel.org/reports/decoupling-natural-resource-use-and-environmental-impacts-economic-growth. Last access: 07 December 2024.
74. E. MacArthur, *Towards the circular economy*, 2013. pp. 23–44. https://www.ellenmacarthurfoundation.org/towards-the-circular-economy-vol-1-an-economic-and-business-rationale-for-an. Last access: 07 December 2024.

75. A. Murray, K. Skene and K. Haynes, The circular economy: An interdisciplinary exploration of the concept and application in a global context. *Journal of Business Ethics*, 2017. 140(3): pp. 369–380. https://doi.org/10.1007/s10551-015-2693-2
76. EMF, *Towards the circular economy, Vol. 1: An economic and business rationale for an accelerated transition*, 2013. https://www.ellenmacarthurfoundation.org/towards-the-circular-economy-vol-1-an-economic-and-business-rationale-for-an. Last access: 01 November 2024.
77. M. Gharfalkar, R. Court, C. Campbell, Z. Ali and G. Hillier, Analysis of waste hierarchy in the European waste directive 2008/98/EC. *Waste Management*, 2015. 39: pp. 305–313. https://doi.org/10.1016/j.wasman.2015.02.007
78. I. Jawahir and R. Bradley, Technological elements of circular economy and the principles of 6R-based closed-loop material flow in sustainable manufacturing. *Procedia Cirp*, 2016. 40: pp. 103–108. https://doi.org/10.1016/j.procir.2016.01.067
79. J. Yan and C. Feng, Sustainable design-oriented product modularity combined with 6R concept: A case study of rotor laboratory bench. *Clean Technologies and Environmental Policy*, 2014. 16(1): pp. 95–109. https://doi.org/10.1007/s10098-013-0597-3
80. J. Potting, M. Hekkert, E. Worrell and A. Hanemaaijer, Circular economy: Measuring innovation in the product chain, in *Planbureau voor de Leefomgeving*, 2017. https://www.pbl.nl/uploads/default/downloads/pbl-2016-circular-economy-measuring-innovation-in-product-chains–2544.pdf. Last access: 07 December 2024.
81. S. Sihvonen and T. Ritola, Conceptualizing ReX for aggregating end-of-life strategies in product development. *Procedia Cirp*, 2015. 29: pp. 639–644. https://doi.org/10.1016/j.procir.2015.01.026
82. Ö. Çimen, Construction and built environment in circular economy: A comprehensive literature review. *Journal of Cleaner Production*, 2021. 305: p. 127180. https://doi.org/10.1016/j.jclepro.2021.127180
83. EC Directive, *Directive (EU) 2018/851 of the European Parliament and of the Council of 30 May 2018 amending directive 2008/98/EC on waste*, 2018. https://eur-lex.europa.eu/legal-content/EN/TXT/?uri=celex%3A32018L0851. Last access: 28 June 2023.
84. P.A. Yost and J.M. Halstead, A methodology for quantifying the volume of construction waste. *Waste Management & Research*, 1996. 14(5): pp. 453–461. https://doi.org/10.1006/wmre.1996.0044
85. H. Yuan and L. Shen, Trend of the research on construction and demolition waste management. *Waste Management*, 2011. 31(4): pp. 670–679. https://doi.org/10.1016/j.wasman.2010.10.030
86. Y.S. Li and X.Q. Zhang, Web-based construction waste estimation system for building construction projects. *Automation in Construction*, 2013. 35: pp. 142–156. https://doi.org/10.1016/j.autcon.2013.05.002
87. Z. Wu, T. Ann, L. Shen and G. Liu, Quantifying construction and demolition waste: An analytical review. *Waste Management*, 2014. 34(9): pp. 1683–1692. https://doi.org/10.1016/j.wasman.2015.12.026
88. ISO, *Environmental management systems – Requirements with guidance for use*, 2015. https://www.iso.org/standard/60857.html. Last access: 28 June 2023.
89. C. Okoli, A guide to conducting a standalone systematic literature review. *Communications of the Association for Information Systems*, 2015. 37(1): p. 43. https://doi.org/10.17705/1CAIS.03743
90. Y. Xiao and M. Watson, Guidance on conducting a systematic literature review. *Journal of Planning Education and Research*, 2019. 39(1): pp. 93–112. https://doi.org/10.1177/0739456X17723971
91. D. Gough, S. Oliver and J. Thomas, *An introduction to systematic reviews*, 2017. Sage. ISBN: 9781473929432.
92. M.J. Grant and A. Booth, A typology of reviews: An analysis of 14 review types and associated methodologies. *Health Information & Libraries Journal*, 2009. 26(2): pp. 91–108. https://doi.org/10.1111/j.1471–1842.2009.00848.x

93. K.S. Khan, R. Kunz, J. Kleijnen and G. Antes, Five steps to conducting a systematic review. *Journal of the Royal Society of Medicine*, 2003. 96(3): pp. 118–121. https://doi.org/10.1258%2Fjrsm.96.3.118
94. A. Aghaei Chadegani, H. Salehi, M. Yunus, H. Farhadi, M. Fooladi, M. Farhadi and N. Ale Ebrahim, A comparison between two main academic literature collections: Web of Science and Scopus databases. *Asian Social Science*, 2013. 9(5): pp. 18–26. https://doi.org/10.5539/ass.v9n5p18
95. R. Jin, H. Yuan and Q. Chen, Science mapping approach to assisting the review of construction and demolition waste management research published between 2009 and 2018. *Resources, Conservation and Recycling*, 2019. 140: pp. 175–188. https://doi.org/10.1016/j.resconrec.2018.09.029
96. F. Jalaei, M. Zoghi and A. Khoshand, Life cycle environmental impact assessment to manage and optimize construction waste using building information modeling (BIM). *International Journal of Construction Management*, 2021. 21(8): pp. 784–801. https://doi.org/10.1080/15623599.2019.1583850
97. R. Quiñones, C. Llatas, M.V. Montes and I. Cortés, A multiplatform BIM-integrated construction waste quantification model during design phase. The case of the structural system in a Spanish building. *Recycling*, 2021. 6(3): p. 62. https://doi.org/10.3390/recycling6030062
98. R. Quiñones, C. Llatas, M.V. Montes and I. Cortés, Quantification of construction waste in early design stages using BIM-based tool. *Recycling*, 2022. 7(5): p. 63. https://doi.org/10.3390/recycling7050063
99. Y. Shi and J. Xu, BIM-based information system for econo-enviro-friendly end-of-life disposal of construction and demolition waste. *Automation in Construction*, 2021. 125: p. 103611. https://doi.org/10.1016/j.autcon.2021.103611
100. J. Wang, H. Wu, H. Duan, G. Zillante, J. Zuo and H. Yuan, Combining life cycle assessment and Building Information Modelling to account for carbon emission of building demolition waste: A case study. *Journal of Cleaner Production*, 2018. 172: pp. 3154–3166. https://doi.org/10.1016/j.jclepro.2017.11.087
101. J.P. Xu, Y. Shi, Y.C. Xie and S.W. Zhao, A BIM-Based construction and demolition waste information management system for greenhouse gas quantification and reduction. *Journal of Cleaner Production*, 2019. 229: pp. 308–324. https://doi.org/10.1016/j.jclepro.2019.04.158
102. B.C. Guerra, F. Leite and K.M. Faust, 4D-BIM to enhance construction waste reuse and recycle planning: Case studies on concrete and drywall waste streams. *Waste Management*, 2020. 116: pp. 79–90. https://doi.org/10.1016/j.wasman.2020.07.035
103. A. Bakchan, K.M. Faust and F. Leite, Seven-dimensional automated construction waste quantification and management framework: Integration with project and site planning. *Resources Conservation and Recycling*, 2019. 146: pp. 462–474. https://doi.org/10.1016/j.resconrec.2019.02.020
104. W. Lu, X. Chen, Y. Peng and L. Shen, Benchmarking construction waste management performance using big data. *Resources, Conservation and Recycling*, 2015. 105: pp. 49–58. https://doi.org/10.1016/j.resconrec.2015.10.013
105. D. Lee, S. Kim and S. Kim, Development of hybrid model for estimating construction waste for multifamily residential buildings using artificial neural networks and ant colony optimization. *Sustainability*, 2016. 8(9). https://doi.org/10.3390/su8090870
106. W.S. Lu, Y. Peng, X. Chen, M. Skitmore and X.L. Zhang, The S-curve for forecasting waste generation in construction projects. *Waste Management*, 2016. 56: pp. 23–34. https://doi.org/10.1016/j.wasman.2016.07.039
107. Y.C. Kim, W.H. Hong, J.W. Park and G.W. Cha, An estimation framework for building information modeling (BIM)-based demolition waste by type. *Waste Management & Research*, 2017. 35(12): pp. 1285–1295. https://doi.org/10.1177/0734242X17736381
108. F. Kleemann, H. Lehner, A. Szczypińska, J. Lederer and J. Fellner, Using change detection data to assess amount and composition of demolition waste from buildings

in Vienna. *Resources, Conservation and Recycling*, 2017. 123: pp. 37–46. https://doi.org/10.1016/j.resconrec.2016.06.010
109. Y.L. Song, Y. Wang, F. Liu and Y.X. Zhang, Development of a hybrid model to predict construction and demolition waste: China as a case study. *Waste Management*, 2017. 59: pp. 350–361. https://doi.org/10.1016/j.wasman.2016.10.009
110. B.C. Guerra, A. Bakchan, F. Leite and K.M. Faust, BIM-based automated construction waste estimation algorithms: The case of concrete and drywall waste streams. *Waste Management*, 2019. 87: pp. 825–832. https://doi.org/10.1016/j.wasman.2019.03.010
111. Z. Wang, H. Li and X. Zhang, Construction waste recycling robot for nails and screws: Computer vision technology and neural network approach. *Automation in Construction*, 2019. 97: pp. 220–228. https://doi.org/10.1016/j.autcon.2018.11.009
112. Z. Wang, H. Li and X. Yang, Vision-based robotic system for on-site construction and demolition waste sorting and recycling. *Journal of Building Engineering*, 2020. 32: p. 101769. https://doi.org/10.1016/j.jobe.2020.101769
113. W. Xiao, J. Yang, H. Fang, J. Zhuang and Y. Ku, Classifying construction and demolition waste by combining spatial and spectral features, in *Proceedings of the institution of civil engineers-waste and resource management*, 2020. Thomas Telford Ltd. https://doi.org/10.1680/jwarm.20.00008
114. G. Coskuner, M.S. Jassim, M. Zontul and S. Karateke, Application of artificial intelligence neural network modeling to predict the generation of domestic, commercial and construction wastes. *Waste Management & Research*, 2021. 39(3): pp. 499–507. https://doi.org/10.1177/0734242X20935181
115. P. Davis, F. Aziz, M.T. Newaz, W. Sher and L. Simon, The classification of construction waste material using a deep convolutional neural network. *Automation in Construction*, 2021. 122: p. 103481. https://doi.org/10.1016/j.autcon.2020.103481
116. R. Hu, K. Chen, W. Chen, Q. Wang and H. Luo, Estimation of construction waste generation based on an improved on-site measurement and SVM-based prediction model: A case of commercial buildings in China. *Waste Management*, 2021. 126: pp. 791–799. https://doi.org/10.1016/j.wasman.2021.04.012
117. W. Lu, J. Lou, C. Webster, F. Xue, Z. Bao and B. Chi, Estimating construction waste generation in the Greater Bay Area, China using machine learning. *Waste Management*, 2021. 134: pp. 78–88. https://doi.org/10.1016/j.wasman.2021.08.012
118. L. Yuan, W. Lu and F. Xue, Estimation of construction waste composition based on bulk density: A big data-probability (BD-P) model. *Journal of Environmental Management*, 2021. 292: p. 112822. https://doi.org/10.1016/j.jenvman.2021.112822
119. X. Chen, H. Huang, Y. Liu, J. Li and M. Liu, Robot for automatic waste sorting on construction sites. *Automation in Construction*, 2022. 141: p. 104387. https://doi.org/10.1016/j.autcon.2022.104387
120. Z. Dong, J. Chen and W. Lu, Computer vision to recognize construction waste compositions: A novel boundary-aware transformer (BAT) model. *Journal of Environmental Management*, 2022. 305: p. 114405. https://doi.org/10.1016/j.jenvman.2021.114405
121. K. Lin, T. Zhou, X. Gao, Z. Li, H. Duan, H. Wu, G. Lu and Y. Zhao, Deep convolutional neural networks for construction and demolition waste classification: VGGNet structures, cyclical learning rate, and knowledge transfer. *Journal of Environmental Management*, 2022. 318: p. 115501. https://doi.org/10.1016/j.jenvman.2022.115501
122. W. Lu, J. Chen and F. Xue, Using computer vision to recognize composition of construction waste mixtures: A semantic segmentation approach. *Resources, Conservation and Recycling*, 2022. 178: p. 106022. https://doi.org/10.1016/j.resconrec.2021.106022
123. S. Na, S. Heo, S. Han, Y. Shin and M. Lee, Development of an artificial intelligence model to recognise construction waste by applying image data augmentation and transfer learning. *Buildings*, 2022. 12(2): p. 175. https://doi.org/10.3390/buildings12020175
124. S. Ray, K.T.W. Ng, T.S. Mahmud, A. Richter and N. Karimi, Temporal analysis of settlement areas and city footprints on construction and demolition waste quantification using Landsat satellite imagery. *Sustainable Cities and Society*, 2024. 105: p. 105351. https://doi.org/10.1016/j.scs.2024.105351

125. S. Ray, K.T.W. Ng, T.S. Mahmud, A. Richter and F. Naghibalsadati, Quantification of construction and demolition waste disposal behaviors during COVID-19 using satellite imagery. *Environmental and Sustainability Indicators*, 2024. 24: p. 100502. https://doi.org/10.1016/j.indic.2024.100502
126. O. Faniran and G. Caban, Minimizing waste on construction project sites. *Engineering, Construction and Architectural Management*, 1998. 5(2): pp. 182–188. https://doi.org/10.1108/eb021073
127. B. Nikmehr, M.R. Hosseini, J. Wang, N. Chileshe and R. Rameezdeen, BIM-based tools for managing construction and demolition waste (CDW): a scoping review. *Sustainability*, 2021. 13(15): p. 8427. https://doi.org/10.3390/su13158427
128. M. Kor, I. Yitmen and S. Alizadehsalehi, An investigation for integration of deep learning and digital twins towards construction 4.0. *Smart and Sustainable Built Environment*, 2023. 12(3): pp. 461–487. https://doi.org/10.1108/SASBE-08-2021-0148
129. Y. Pan and L. Zhang, Roles of artificial intelligence in construction engineering and management: A critical review and future trends. *Automation in Construction*, 2021. 122: p. 103517. https://doi.org/10.1016/j.autcon.2020.103517
130. R. Sacks, M. Girolami and I. Brilakis, Building information modelling, artificial intelligence and construction tech. *Developments in the Built Environment*, 2020. 4: p. 100011. https://doi.org/10.1016/j.dibe.2020.100011
131. T.O. Ayodele, Types of machine learning algorithms. *New Advances in Machine Learning*, 2010. 3: pp. 19–48. https://doi.org/10.5772/225
132. W. Lu, Y. Peng, X. Chen, M. Skitmore and X. Zhang, The S-curve for forecasting waste generation in construction projects. *Waste Management*, 2016. 56: pp. 23–34. https://doi.org/10.1016/j.wasman.2016.07.039
133. Y. Ku, J. Yang, H. Fang, W. Xiao and J. Zhuang, Deep learning of grasping detection for a robot used in sorting construction and demolition waste. *Journal of Material Cycles and Waste Management*, 2021. 23: pp. 84–95. https://doi.org/10.1007/s10163-020-01098-z
134. J. Chen, W. Lu, L. Yuan, Y. Wu and F. Xue, Estimating construction waste truck payload volume using monocular vision. *Resources, Conservation and Recycling*, 2022. 177: p. 106013. https://doi.org/10.1016/j.resconrec.2021.106013
135. S.O. Ajayi, L.O. Oyedele, M. Bilal, O.O. Akinade, H.A. Alaka and H.A. Owolabi, Critical management practices influencing on-site waste minimization in construction projects. *Waste Management*, 2017. 59: pp. 330–339. https://doi.org/10.1016/j.wasman.2016.10.040
136. C.S. Poon, T. Ann and L. Ng, On-site sorting of construction and demolition waste in Hong Kong. *Resources, Conservation and Recycling*, 2001. 32(2): pp. 157–172. https://doi.org/10.1016/S0921–3449(01)00052–0
137. W. Xiao, J. Yang, H. Fang, J. Zhuang, Y. Ku and X. Zhang, Development of an automatic sorting robot for construction and demolition waste. *Clean Technologies and Environmental Policy*, 2020. 22: pp. 1829–1841. https://doi.org/10.1007/s10098-020-01922-y
138. J. Chen, Y. Fu, W. Lu and Y. Pan, Augmented reality-enabled human-robot collaboration to balance construction waste sorting efficiency and occupational safety and health. *Journal of Environmental Management*, 2023. 348: p. 119341. https://doi.org/10.1016/j.jenvman.2023.119341
139. H. Wu, J. Wang, H. Duan, L. Ouyang, W. Huang and J. Zuo, An innovative approach to managing demolition waste via GIS (geographic information system): A case study in Shenzhen city, China. *Journal of Cleaner Production*, 2016. 112: pp. 494–503. https://doi.org/10.1016/j.jclepro.2015.08.096
140. B. Yu, J.Y. Wang, J. Li, J.R. Zhang, Y.N. Lai and X.X. Xu, Prediction of large-scale demolition waste generation during urban renewal: A hybrid trilogy method. *Waste Management*, 2019. 89: pp. 1–9. https://doi.org/10.1016/j.wasman.2019.03.063
141. B. Yu, J. Wang, J. Li, J. Zhang, Y. Lai and X. Xu, Prediction of large-scale demolition waste generation during urban renewal: A hybrid trilogy method. *Waste Management*, 2019. 89: pp. 1–9. https://doi.org/10.1016/j.wasman.2019.03.063
142. Y. Jiao, Y. Wang, S. Zhang, Y. Li, B. Yang and L. Yuan, A cloud approach to unified lifecycle data management in architecture, engineering, construction and facilities

management: Integrating BIMs and SNS. *Advanced Engineering Informatics*, 2013. 27(2): pp. 173–188. https://doi.org/10.1016/j.aei.2012.11.006
143. T. Bock, The future of construction automation: Technological disruption and the upcoming ubiquity of robotics. *Automation in Construction*, 2015. 59: pp. 113–121. https://doi.org/10.1016/j.autcon.2015.07.022
144. R.M. Gavilan and L.E. Bernold, Source evaluation of solid waste in building construction. *Journal of Construction Engineering and Management*, 1994. 120(3): pp. 536–552. https://doi.org/10.1061/(ASCE)0733–9364(1994)120:3(536)
145. Q.K. Wang, L. Chen, R.B. Hu, Z.G. Ren, Y.T. He, D.R. Liu and Z.Q. Zhou, An empirical study on waste generation rates at different stages of construction projects in China. *Waste Management & Research*, 2020. 38(4): pp. 433–443. https://doi.org/10.1177/0734242X19886635
146. A. Bakshan, I. Srour, G. Chehab and M. El-Fadel, A field based methodology for estimating waste generation rates at various stages of construction projects. *Resources, Conservation and Recycling*, 2015. 100: pp. 70–80. https://doi.org/10.1016/j.resconrec.2015.04.002
147. A. Vilventhan, V. Ram and S. Sugumaran, Value stream mapping for identification and assessment of material waste in construction: A case study. *Waste Management & Research*, 2019. 37(8): pp. 815–825. https://doi.org/10.1177/0734242X19855429
148. V. Ram and S.N. Kalidindi, Estimation of construction and demolition waste using waste generation rates in Chennai, India. *Waste Management & Research*, 2017. 35(6): pp. 610–617. https://doi.org/10.1177/0734242X17693297
149. P.T. Lam, T. Ann, Z. Wu and C.S. Poon, Methodology for upstream estimation of construction waste for new building projects. *Journal of Cleaner Production*, 2019. 230: pp. 1003–1012. https://doi.org/10.1016/j.jclepro.2019.04.183
150. V.W. Tam, Y. Zhou, C. Illankoon and K.N. Le, A critical review on BIM and LCA integration using the ISO 14040 framework. *Building and Environment*, 2022. 213: p. 108865. https://doi.org/10.1016/j.buildenv.2022.108865
151. ISO, *Environmental management – Life cycle assessment – Principles and framework*, 2006. https://www.iso.org/obp/ui/#iso:std:iso:14040:ed-2:v1:en. Last access: 08 December 2024.
152. M. Malia, J. de Brito, M.D. Pinheiro and M. Bravo, Construction and demolition waste indicators. *Waste Management & Research*, 2013. 31(3): pp. 241–255. https://doi.org/10.1177/0734242X12471707
153. UKGBC (UK Green Building Council), *The importance of retrofitting in advancing net zero*, 2021. https://www.ukgbc.org/news/the-importance-of-retrofitting-in-advancing-net-zero/ Last access: 25 Feb 2023.
154. M. Honic, I. Kovacic and H. Rechberger, Improving the recycling potential of buildings through Material Passports (MP): An Austrian case study. *Journal of Cleaner Production*, 2019. 217: pp. 787–797. https://doi.org/10.1016/j.jclepro.2019.01.212
155. A. Stephan and A. Athanassiadis, Towards a more circular construction sector: Estimating and spatialising current and future non-structural material replacement flows to maintain urban building stocks. *Resources, Conservation and Recycling*, 2018. 129: pp. 248–262. https://doi.org/10.1016/j.resconrec.2017.09.022
156. A.D. Baez, P.V. Saez, M.D. Merino and J.G. Navarro, Methodology for quantification of waste generated in Spanish railway construction works. *Waste Management*, 2012. 32(5): pp. 920–924. https://doi.org/10.1016/j.wasman.2012.01.007
157. K. Cochran and T.G. Townsend, Estimating construction and demolition debris generation using a materials flow analysis approach. *Waste Management*, 2010. 30(11): pp. 2247–2254. https://doi.org/10.1016/j.wasman.2010.04.008
158. R.F. de Magalhães, Â.d.M.F. Danilevicz and T.A. Saurin, Reducing construction waste: A study of urban infrastructure projects. *Waste Management*, 2017. 67: pp. 265–277. https://doi.org/10.1016/j.wasman.2017.05.025

3 Building Information Modelling, Semantic Web Technology, and Life Cycle Assessment

3.1 Introduction

BIM and SWT are the two main technologies supporting the research to enhance the interoperability and automation of the CWQ process. The features of BIM, including its life-cycle information management, automated quantity estimation, parametric modelling, interoperability with external applications, and promoting a collaborative environment through integrated decision-making, primarily facilitated the CWQ process. In addition to enhancing the integration of CW data gathered from distributed data sources, a common data model that is interoperable, open, and easy for machines and humans to communicate with is crucial, which is facilitated by the SWT. SWT, an extension of the traditional web, is integrated into the research to enable a unified CW data structure, seamless communication across various CW platforms, and an interoperable and automated flow of information between BIM and CWQ platforms. The ontology modelled using the SWT functions ensures that the data within the CWQ model is interoperable and standardised. Primarily, the standard data representation and exchange, logical semantic inference, and data integration with BIM are the main attributes of SWT that support the development of a CWQ workflow. Besides these technologies, the overall quantification process adopts the LCA methodology to enable a granular CW quantification and classification. The material life cycle classification and a systematic process emphasised in the LCA framework are reflected in the research. Emphasising the focus on the whole life cycle of materials, enabling comprehensive environmental analysis, and supporting data-driven decision-making to foster CE adoption and whole life-cycle impact reduction are the main rationales for adopting the LCA methodology into CWQ workflows.

3.2 Building Information Modelling

BIM is one of the main technologies that supported the research in proposing a comprehensive CWQ workflow. This section discusses the definitions, components, and benefits of using BIM. It also discusses the maturity levels and dimensions of BIM, the type of BIM-based data models, modelling environments, and the BIM attributes that support the CWQ process.

DOI: 10.1201/9781003610762-3

3.2.1 Definition, Components, and Benefits

BIM was introduced as solid modelling in the late 1970s and early 1980s to overcome the complexities, ambiguities, and fragmentation in traditional construction practices. With continuous improvements in its functionalities, BIM is now called an intelligent modelling tool [1]. It is the process that aids in generating and managing information about a building throughout its life cycle. The output produced by BIM is called the Building Information Model, a data-rich, object-oriented, parametric, and intelligent digital representation of the built asset [1], which characterises the existence of geometrical and non-geometrical information that can be utilised across the project life-cycle. Similarly, the International Organisation for Standardization (ISO) 19650–1:2018 defines BIM as the 'use of shared digital representation of a built asset to facilitate planning, designing, construction and operation process to form a reliable basis for decisions' [2] – P5.

Besides, BIM is a process that makes a complete paradigm shift in the project workflow and delivery methods [3, 4]. The development of BIM applications is perceived as a platform to enhance integrated project delivery methods, collaboration, and communication across the supply chain [1]. It is also highlighted that integrating integrated project delivery with BIM improves the overall project performance [5–7]. Based on the above definitions, it is clear that BIM is not just a 3D modelling tool but also a collaborative process that significantly changes the project workflow to enhance overall efficiency. Inclusive of all, the potential benefits of BIM across the life-cycle stages are driven by (i) technology (3D-nD modelling and simulation, federated BIM model, parametric modelling, clash detection, and visualisation), (ii) information (functional and non-functional information, interoperable Industry Foundation Classes [IFCs] data, life-cycle data, common data environment [CDE], and information management), and (iii) process management (cloud-based management, real-time collaboration, stakeholder integration, seamless information exchange) factors [8, 9].

3.2.2 BIM Maturity Levels

BIM maturity levels are established to demonstrate the degree and type of information, modelling techniques, information exchange methods, and level of collaboration between the project stakeholders. The widely acknowledged BIM maturity model developed by [10] consists of four levels:

Level 0 BIM represents traditional paper-based 2D drafting and information-sharing techniques with zero collaboration among project participants. With advanced digital techniques, this level has become obsolete and is scarcely used by the industry.

Level 1 BIM – Advanced from the previous level, Level 1 incorporated 2D and 3D modelling for product information and conceptual works. A CDE is employed to share information electronically as a model between the stakeholders. However, neither zero nor low collaboration is anticipated from this level.

Level 2 BIM – At this level, the designers are expected to work in an individual 3D model and share information using open data exchange formats like IFC and Construction Operations Building Information Exchange (COBie). The BIM dimensions go far beyond 3D and integrate 4D and 5D data into the model. This information exchange allows several internal and external data to be integrated into a model, thus creating a federated BIM model. Level 2 BIM is a distinguished collaborative level which emphasises the need for streamlined coordination and information exchange across the whole process. The UK Government has mandated the implementation of Level 2 BIM for all publicly procured projects since 2016 [11].

Level 3 BIM promises high stakeholder collaboration, often called 'integrated or open BIM'. Despite other levels, Level 3 allows the project participants to work in a single shared model stored in a central repository. This level is expected to reduce information errors, allowing all the stakeholders to work on the same model simultaneously. Though Level 3 is not widely established, the UK Government's strategic plan outlines the vision for its adoption in the AECO industry between 2016 and 2025 [11].

3.2.3 BIM Level of Model Definition

The level of model definition (LOMD) represents the level of information existing within the developed BIM model to maintain data quality and ease information exchange. The two components of LOMD defined by BS ISO 19650 (replaced version of PAS 1192–2) include the level of model detail (LOD) and level of model information (LOI). The LOD describes the level of graphical data, and the LOI relates to the level of non-graphical information within the BIM model.

Although there is no straightforward relationship between the LOMD and project life-cycle stages, fundamentals highlighted in PAS 1192–2 should be adopted to ease information flows across different parties. The LOMD consists of seven

Table 3.1 LOMD and Life Cycle Stages

LOMD	*Life-cycle stages*	*RIBA plan of work 2020*	*AIA's LOD*
1	Brief	Preparation and Brief	–
2	Concept	Conceptual Design	LOD 100
3	Definition	Spatial Coordination	LOD 200
4	Design	Technical Design	LOD 300
5	Build and Commission	Manufacturing and Construction	LOD 400
6	Handover and close-out (As built BIM)	Handover	
7	Operation and in-use (AIM)	Use	LOD 500

Source: Drafted by the author based on descriptions provided in the literature levels, from LOMD 1 to LOMD 7. Likewise, the American Institute of Architects (AIA) publishes an LOD framework, calling LOD a 'Level of Development' rather than a 'Level of Detail'. Here, the LOD consists of five stages ranging from LOD 100 to LOD 500. In both cases, the quantity and accuracy of information increase as the levels increase. Table 3.1 presents the relationship between LOMD, LOD, project-life-cycle stages and RIBA plan of work stages 2020. The BIM-based semantic CWQ framework is designed to meet the LOMD 3 requirements for integrating the CW analysis with actual project workflows.

3.2.4 BIM Dimensions

The type of information a BIM model holds decides its dimensions. The dimensions of BIM refer to the uses of BIM across various activities of the project life cycle. In the early days, the 3D capacity of BIM was the primary purpose for its adoption. However, additional data was added to the existing BIM model to meet the project requirements and make timely decisions. Integrating additional data led to the expansion of BIM from 3D to nD [12, 13]. The main dimensions of BIM discussed in the literature include the following.

- **3D BIM** – It is the primary layer of BIM. The 3D BIM contains the essential 3D geometric representation of the building components that can be used to review and visualise designs.
- **4D BIM** – The addition of time/scheduling data to 3D BIM is called 4D BIM. It helps the stakeholders visualise the construction progress, manage project timelines, and coordinate activities to avoid clashes during their delivery [12].
- **5D BIM** – Integrating cost estimates into 3D is considered the fifth dimension. The main purpose of 5D is cost estimation, resource allocation, and visualising real-time project changes on cost and budgeting for efficient construction management [13].
- **6D BIM** – The research studies [14–16] attribute 6D to the addition of sustainability information. However, this stage is still in the infant stage and requires precise evidence of what information is included in the BIM model to avoid inconsistencies and enhance transparency [13, 17].
- **7D BIM** – The literature claims that integrating facilities management information into the 3D BIM is attributed to 7D. However, according to [18] and other research studies [19, 20], integrating facilities management data into the BIM is called 6D and 7D is attributed to sustainability information. For this research, integrating sustainability data with 3D is considered 6D BIM.

Although there are multiple dimensions, only the 3D, 4D, and 5D BIM dimensions are widely agreed upon, and there is a lack of consensus beyond the fifth dimension [17]. In this research, the quantified CW data is integrated into the 3D BIM. Since CW generation is one of the leading indicators influencing environmental, economic, and social sustainability, the research output produces a 6D BIM that facilitates CWQ.

3.2.5 BIM Data Models

The BIM model is a digital output of the BIM process that consists of a broad range of information about the built asset. The data and information exchanged among different systems and participants are crucial to achieving a collaborative environment. Depending on the interoperability and accessibility, the standard data models of BIM include IFC, COBie, and Green Building XML (gbXML). They are called open-source non-proprietary data models, facilitating the AECO industry to overcome interoperability issues.

3.3.5.1 Industry Foundation Classes

IFC is a standardised open data model that exchanges information within multiple AECO domain. It is one of the distinguished schemas deployed by buildingSMART to ease data flows and increase the efficiency of the information exchange in a fragmented, data-intensive AECO industry. The IFC model presents a comprehensive, object-oriented, and universal data model of the project [21]. It contains design and construction information, including geometries, material types, material properties, and relationships between them. The desired IFC can be encoded in various formats, including STEP (SPF), Extensible Markup Language (XML), JavaScript Object Notation (JSON), Turtle, and Resource Description Framework (RDF) [22]. Each format has unique advantages; however, the STEP (IFC-SPF) format is widely used for its readability and compactness. It is based on the EXPRESS schema defined by international standard ISO 10303–21. The other file formats, like JSON and XML, provide enhanced readability and have a broad range of applications across multiple software platforms. Nevertheless, the file size is three to four times larger than the STEP file and is not widely used [22]. Besides, the Turtle and RDF formats support the semantic web and linked data applications of IFC models using Web Ontology Language (OWL), called ifcOWL [23].

3.3.5.2 Construction Operations Building Information Exchange

COBie defines asset information rather than building geometries. It consists of structured and consistent information that supports post-occupancy decision-making [24]. COBie is typically a spreadsheet of asset information that eases information exchange between the construction and operation phases [25]. The COBie worksheet includes information from the initial design to the handover phase, including product details, manufacturer, warranties, service life, and maintenance schedule. COBie is a real-time digital inventory that supports operation, maintenance, and asset management, switching from traditional paper-based information management. This data further supports design, construction, scheduling, environmental analysis, and costs during the in-use stages [24, 26]. However, it is more likely to be used in the operation and maintenance stages, thus having minimal application to the research.

3.3.5.3 Green Building XML (gbXML)

gbXML supports exchanging information stored in BIM models to various building analysis platforms, thus enabling interoperability between them [27]. The gbXML was introduced by Green Building Studio in 2000 to facilitate energy analysis of BIM models. The file format, consisting of multiple elements and attributes, is widely used for sustainable design and energy analysis and is supported by several BIM vendors. Moreover, this schema can import and export over 50 engineering and analysis tools, thus facilitating the designers and engineers to develop energy-efficient sustainable buildings. However, the format has minimal application in quantifying CW.

Among the discussed non-proprietary BIM data models, the attributes of ifcOWL hold higher chances for developing the semantic CW data model. However, previous studies [23, 28] emphasise that the heavy volume of information and lack of specific standard rules of the ifcOWL model consume more time and lead to various complexities during its processing. So, considering these limitations, the research prefers to work with proprietary data models and expand the functionalities later to align with open data models.

3.2.6 BIM Modelling Environments

Some proprietary data models exist apart from standard data models. The ones created by the specific software developer are executable only in the authorised software. Some widely used BIM software are Autodesk Revit, ArchiCAD, Vectorworks Designer, Tekla structures, MicroStation, Allplan, Solidworks, and SketchUp Pro. Among them, Autodesk Revit was selected in the research to establish its integration with the semantic data model. The reasons for including Autodesk Revit include its wide-ranging applications, API features, interoperability with external models/platforms, data model supports, visualisation and information richness, and parametric modelling features. Besides, the National Building Specification (NBS) report highlights that around 50% of AECO professionals prefer Autodesk Revit to develop BIM models [29]. In addition, the .NETAPI and Dynamo Revit led to the seamless integration of external tools, plugins, and software with 3D BIM models. Hence, considering these features, the Autodesk Revit 2022 version is selected for the research.

3.3 BIM Features Facilitating CWQ

BIM, one of the leading object-oriented collaborative technologies, is accounted for in this research. The inherent features of BIM that support CWQ across the life-cycle stages, from conceptual design to EoL, are discussed in this section. It includes the following: (i) life-cycle information management, (ii) automated quantity estimation, (iii) parametric modelling, (iv) interoperability with external applications, and (v) promoting a collaborative environment.

3.3.1 Life Cycle Information Management

BIM models provide piles of functional and non-functional information about a project that can be stored, accessed, managed, exchanged, and reused throughout the project life cycle [8, 30]. The multi-dimensional capability of BIM characterises project information, including design details, MTOs, schedule plans, cost reports, sustainability performance, and operation and maintenance information (discussed in Section 3.2.4). This diverse information associated with a project is assembled into one single model, named federated BIM or combined BIM model. The federated BIM, compiled from various disciplines, enhances information flows and management across the life-cycle stages. Besides, working on a single shared model provides sufficient information for the stakeholders to identify errors, detect clashes, design simulations, allocate resources, and make informed decisions in a virtual environment [8]. It also eases communication and collaboration among the project participants, thus obtaining the benefits of integrated project deliveries [1]. So, CW being an output of various project factors, the BIM model would provide essential information required to assess and optimise their generation across the life-cycle stages.

3.3.2 Automated Quantity Estimation

Traditionally, the material details are estimated manually with 2D drawings provided by the designers. As complex entities, buildings and infrastructures involve diverse materials across the project stages that must be appropriately accounted for in quantity estimation. The estimation enabled via this conventional method is time-consuming and requires high human intervention. These drawbacks are now resolved by BIM-based MTOs, allowing the user to automatically extract the geometric and material properties of building elements used in the design [1]. Since the accuracy of CWQ is highly dependent on the quantity and quality of input quantities, the BIM-based MTO is utilised to mitigate data errors and enhance the granularity of waste results [31, 32]. These reports will, therefore, increase the chances of optimisation, resource allocation, waste storage, and planning from the early design stages.

3.3.3 Parametric Modelling

Buildings are complex entities containing several interconnected elements that undergo different functions across the project life cycle. So, making changes in a specific element influences all its adjacent elements and its properties, thus requiring more time and manual work to update the changes. This drawback is resolved by parametric modelling, a dynamic process with functional intelligence capability that would update the changes automatically according to the pre-established relationships of building elements [30, 33]. Parametric modelling is a critical theme that runs across the spine of BIM. Therefore, quantifying, optimising, and reporting CW as a design parameter for the selected element simultaneously updates the whole family. Besides, BIM-based parametric modelling allows designers to interact with the CW platforms and optimise designs within a limited time and effort. This further improves the visualisation and automation of the whole system, enabling the project team to develop comprehensive CWM plans.

3.3.4 Interoperability With External Applications

Despite the modelling advancements achieved by BIM, interoperability is still considered a major drawback for its adoption among AECO professionals. For instance, the survey conducted by McGraw Hill Construction [34] proved that improving interoperability between software applications could significantly improve the business value of BIM. The interoperability also hinders BIM applications in the domain of environmental and performance analysis [35, 36]. In the current scenario, the functionality of BIM authoring tools is expanded to communicate with external applications through APIs and plugins.

The research conducted by the authors [37, 38] used the API features of Autodesk Revit to enhance interaction between BIM and CWQ platforms. However, these models are designed to meet specific BIM software requirements and lack interoperability with other BIM applications. Though open-data formats like IFC, XML, and OWL offer unified data structure and high interoperability, their application in the CW aspect is scant. This gap compels the need for critical analysis of how open-data formats could benefit information exchange between BIM and CW platforms. The focus should also be placed on enhancing the bidirectional flow of information, meaning that changes in one programme should seamlessly flow across the connected programmes.

3.3.5 Promoting Collaborative Environment

Collaboration and communication are critical factors determining the project's success [39, 40]. Besides traditional fragmented processes, BIM offers a collaborative virtual environment encompassing various systems and disciplines within a single digital model to enhance communication and decision-making [41, 42]. The quality, transparency, and accessibility of information in the BIM model improve stakeholder engagement, further leading to informed decision-making. The same scenario applies to CW, where multi-disciplinary collaboration from the early design stages is vital for making competent WM plans [43]. Thus, the technology (seamless information flows) and process-based collaboration (stakeholder engagement) in CWM could be attained through BIM implementation. Furthermore, integrating CW workflows with BIM would significantly enhance their accessibility and applicability across the life cycle stages and support users in making informed decisions.

3.4 Semantic Web Technology

In addition to BIM, SWT played a vital role in developing an interoperable and automated CWQ framework. This section discusses the definitions, components, and benefits of integrating SWT to develop an ontology. The modelling language and the engineering method selected to model an ontology are also added to the discussion. The benefits of integrating SWT with BIM and the functionalities of SWT enhancing the CWQ process are discussed.

3.4.1 Definition, Concepts, and Functionalities

The concept of the semantic web was coined by Tim-Berners-Lee in 1999 to overcome the limitations of the traditional World Wide Web and make meaningful information accessible automatically across multiple information systems [44]. The semantic web, a component of Web 3.0, is not different from the traditional Web but an extension of WWW running with international standards designed by the World Wide Web Consortium (W3C). The W3C identifies semantic web as 'a common framework that allows data to be shared and reused across application, enterprise, and community boundaries' [45] – P1. The semantic web enables a shift from the traditional web from being a 'document medium for people' to a 'data and information medium that can be exchanged and integrated automatically' [44, 46]. Besides, the core part of this web design is that the semantic information should be readable, integrated, and exchanged by humans and machines. So, the data from the web (called the 'web of data') that is machine-readable and actionable is called the 'semantic web' [44]. It accesses information based on the 'meaning' rather than the 'syntax', allowing machines to recognise data and extract knowledge across various sources.

The growing technologies generate piles of heterogeneous data that must be seamlessly integrated to discover patterns and acquire knowledge [46]. These gaps are filled by the semantic web, which allows machines to analyse data independently and have seamless interaction with various systems, applications, and disciplines [44]. Apart from the traditional web, the data retrieved from the semantic web are more accurate, interoperable, and consistent. The logical relations between the interconnected terms enable interoperability between different systems [46]. Also, the meanings are derived from a structured data hierarchy to develop knowledge domains. Structured information that can enhance reasoning and knowledge representation is considered to facilitate semantic web functions. Overall, the versatility and the ability to store and process metadata call the semantic web a robust decentralised system that enables efficient collaboration between machines and humans [44], which is critical to enhancing the productivity and functionalities of CWQ platforms.

3.4.2 Semantic Web Architecture

The semantic web is developed by a layered approach, where each step builds a layer on top of the previous layer. Tim-Berners-Lee proposed this layered approach. The layers are called semantic web stack or a layer cake, which illustrates the architecture of semantic web development [47]. Figure 3.1 shows the main layers of the semantic web stack. The stack development is still evolving, and the latest version was proposed by [48]. The stack shows that the semantic web is not a replacement for the traditional web but its extension.

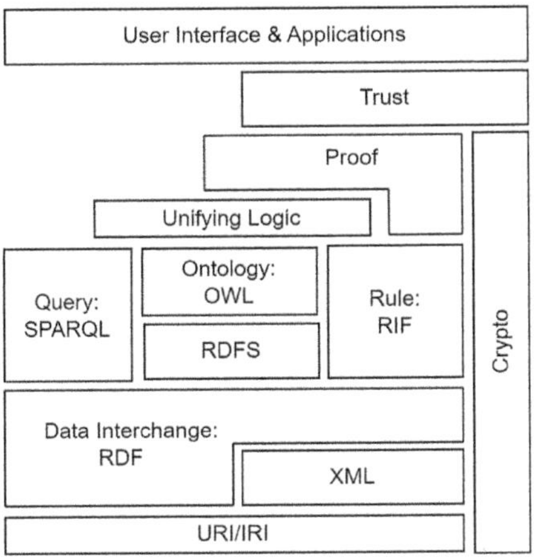

Figure 3.1 Semantic Web Stack

Source: Adapted from [47]

3.4.2.1 *Bottom Layers*

i) *The reference layer (Unicode/URI/IRI)* is the bottom and foundation layer of the semantic web. Unicode is a character encoding system used in software applications to encode, represent, handle, and exchange data across multiple information technology platforms [49]. The Unique Resource Identifier (URI) is a sequence of characters that helps to identify a specific resource, and an Internationalised Resource Identifier (IRI) allows encoding of the web address with Unicode [50].

ii) *The syntax layer (XML)* is the second layer of the stack, which provides a user-friendly vocabulary to structure and share documents across the web [47].

3.4.2.2 *Middle Layers*

i) *RDF* is a basic framework that represents the relationship between entities in a structured form to enhance machine readability. The RDF will be represented as triples (subject, predicate, and object) in a graphical format. The syntax of RDF can be written in multiple forms, including N3, Turtle, and RDF/XML JSON-LD [47].

ii) *RDF schema (RDFS)* is a schema that provides basic modelling primitive language to order web objects into hierarchies. The primitives include classes, sub-classes, sub-property relationships, domains, and range limits. RDF schema is based on RDF. It can be seen as a primitive language for ontology development but lacks applications to establish comprehensive semantics between entities [47].

iii) *OWL* expands the RDFS language by adding more constructs to describe the complex semantics of RDF statements. It allows additional constraints such as cardinality, restrictions of values, and transitivity. Since OWL depends on the description logic, the reasoning power of the semantic web is significantly enhanced [47].

iv) *Rule interchange format* defines a standard for exchanging rules among rule systems. It is used in scenarios where the entity relationships cannot be defined using the description logic used in OWL [47, 51].

v) *SPARQL* is a query language for RDF triples. It enables users to query information from RDF databases or view data as RDF using middleware [47].

3.4.2.3 Top Layers

i) *The logic layer* enhances ontology language and allows application-specific descriptive knowledge [47].

ii) *The proof layer* executes the rules and represents the proofs in web languages and proof validation [47].

iii) *The trust layer* – Based on the recommended trusted bodies, the digital signatures of RDF statements will be used to verify if the data comes from a trusted source [47].

iv) *The UI* is the final layer that enables users to access semantic web applications.

3.4.3 Ontology – Definitions

Ontology, the backbone of the semantic web, is defined as explicit specifications of concepts, objects, and other entities in a specific domain and their relations. Ontologies are metadata schemas that provide consistent vocabularies, each with explicit specification and machine-readable semantics [52]. From the philosophical stance, ontology describes the nature of beings and their relations [53]. In the computational field, an ontology defines standard semantics to ease understanding of the information and knowledge sharing within a domain by conceptualising the entities and relations among them [54]. There are multiple definitions of ontologies. The one mostly cited defines ontology as *an* 'explicit specification of conceptualization' [53]. Expanding this definition, Noy and McGuinness [54] – P3 – defines ontology as *an*

> a formal, explicit description of concepts in a domain of discourse (classes [sometimes called concepts]), properties of each concept describing various features and attributes of the concept (slots [sometimes called roles or properties]), and restrictions on slots (facets [sometimes called role restrictions]). An ontology with a set of individual instances of classes constitutes a knowledge base.

Although no standard definition of ontologies exists, computational ontology aims to increase knowledge sharing and interoperability across multiple systems and disciplines with standard entities and define explicit relationships between them as prerequisites. Moreover, the ontology should be formal, shareable, extendable, and reusable.

3.4.4 Components of an Ontology

The main components of an ontology [53–55] include the following:

i) *Classes* represent a set of entities or things within the selected domain. They are also called 'concepts'.
    a) *Primitive classes* are entities whose properties have only the necessary condition(s) to be a class member.
    b) *Defined classes* are entities whose properties have both necessary and sufficient condition(s) to be a class member.
ii) *Instances* are individual things represented by a concept. They are the basic ground-level objects.
iii) *Attributes* are the properties, features, parameters, or aspects of objects and classes.
iv) *Relations* represent the relationships between objects in an ontology. The set of relations describes the semantic relationships of the domain.
v) *Restrictions* are formal descriptions that provide the criteria for input to be a member of the classes.
vi) *Axioms* are used to constrain the values for instances or classes.

3.4.5 Ontology Modelling Languages and Frameworks

An ontology must be encoded in predefined formal languages to enhance the reasoning mechanism throughout the ontology lifecycle. Several logic and web-based standard languages are used for ontology modelling based on the semantic architecture/stack. The languages include OIL, DAML, DAML+OIL, SHOE, SHACL, and OWL [55–57]. However, the RDF and its extension languages, RDFS and OWL, act as standard languages for representing semantic data (Allemang and Hendler, 2011), which are widely recommended by W3C [56].

3.4.5.1 Resource Description Framework (RDF)

RDF is a standard framework for representing web-based semantic data. It is a basic framework used to model the meta-data and the relationship between entities in a structured form to enhance machine readability rather than being processed only by humans. In the current scenario, incorporating diverse technologies leads to the generation of big data that needs to be seamlessly integrated across multiple applications. Big data represents the data's large volume, variety, and velocity, called 3Vs [58]. This is the typical scenario in the AECO industry. The heterogeneity (variety) being a predominant barrier to achieving interoperability, the RDF provides a common framework that allows fragmented information to be linked and exchanged with multiple systems, applications, and disciplines on the web [59]. The output of the RDF model will be linked data that has enhanced interoperability and applications.

The RDF data will be represented in an easy-to-understand labelled graphical format called triples. The triples include subject, object, and predicate, in which the 'subject' and 'object' are linked via a 'predicate'. Hence, the RDF data model is called a triple store. It creates a chain of triples, where an object(s) of one triple would form a subject(s) of another connected through predicate(s). Furthermore, predicates that have literal values are called data properties. Besides, RDF uses URIs to identify information and describe the entities and their property values with reduced ambiguities. It enables the designer to create links between entities, which is the core value of RDF. Thus, the subject is a URI, the predicate is a resource, so it is a URI, and the object is either a URI or a literal value [56].

3.4.5.2 Web Ontology Language (OWL)

OWL is a vocabulary extension of RDF; thus, any RDF graphs form an OWL full ontology. Like RDF, it is a W3C-recommended language, and it is derived from the DAML+OIL language. OWL is introduced to process the information rather than just presenting the contents to the user. It is a computational logic-based language that supports greater machine interoperability than other schemas like XML and RDF. It adds more vocabulary for describing classes and properties compared to other schemas. OWL has three increasingly expressive sub-languages: OWL-Lite, OWL-Description Logic (OWL-DL) and OWL-Full [60]. However, the RDF documents will typically be built in OWL full unless specifically constructed in OWL-DL or OWL Lite [61].

The latest version of OWL is OWL 2, an extension of the earlier version of OWL 1 developed to enhance information accessibility and interoperability. The different profiles/sub-languages of OWL 2 ontology are OWL 2 existential quantification (EL), OWL 2 query language (QL) and OWL 2 rule language (RL). The OWL 2 EL is used to employ ontologies that contain many classes and properties. It captures the expressive power of the ontologies, and fundamental reasoning problems can be performed. The OWL 2 QL is applied where many instance data and query answering are critical reasoning tasks. The OWL 2 RL is used in applications that require scalable reasoning without using excessive power. It can be implemented in rule-based reasoning engines. The entities, classes, properties, axioms, and restrictions supported by these profiles vary with each other. The OWL 2 ontologies store and exchange linked data using concrete syntaxes like RDF/XML, Turtle, and XML to integrate across multiple applications. However, the primary exchange syntax used is RDF/XML for its computability across all OWL2 tools [62], which is used in this research to store and query ontologies.

3.4.5.3 SPARQL Protocol and RDF Query Language

The standard way to access the ontology data is through query languages, the next semantic web stack layer. Several languages, including SPARQL Protocol and RDF Query Language (SPARQL), semantic web rule language, RDF query language, and Sesame RDF query language, are used to query knowledge from the ontologies [63–65]. Among all existing languages, SPARQL (version 1.1) is selected to query the ontology designed in this research. The critical reason for selecting SPARQL is that it is one of the W3C-acknowledged languages to query semantic web ontologies presented in RDF syntax. SPARQL aims to build queries to extract, analyse, and manipulate RDF data models stored on the web or in an RDF store. SPARQL is a 'tell and ask' system that presents queries as statements. It has structured query language syntax, and the query patterns are represented as triples [56, 66]. Besides, SPARQL use expressive queries that enable flexible and detailed information retrieval from RDF data. Designed to query a diverse set of data, SPARQL can extract information from distributed datasets that can be mapped to RDF. The significant shortcomings of SPARQL are that it works closely with RDF syntax itself and lacks a native understanding of OWL. However, using SPARQL is appropriate because the RDF/XML syntax is primarily used in this research to store and exchange ontologies across multiple applications. This syntax is preferred to ease the integration of ontologies with BIM platforms (discussed in Section 6.2.1).

3.4.6 Ontology Modelling Environment

Ontologies are independent of the environment used to model them, thus leading to knowledge maintenance and interoperability. Some ontology environments include Protégé, NeOn Toolkit, Onto Studio, Vitro, Swoop, Isa Viz, Apollo, Onto Studio, Owl Sight, VisualKii, and VisualKnowledge [67]. Among these environments, Protégé was selected to model the ontology in this research. It is a free, open-source tool developed by the Stanford Centre for Biomedical Information Research with tools to construct domain ontology and develop a knowledge base environment.

The critical features accounted for selecting the Protégé include its explicit benefits, including its ontology editor, interoperability, and collaboration with other environments, extensive architecture, extensibility using plugins, the usability of tools, and customised UI for ontology modelling and visualisation features [68–70]. Furthermore, this editor allows data to be imported and exported in multiple formats. It also provides enhanced usability for merging and reusing existing ontologies. The next noticeable feature is its extensibility via APIs, making it flexible for rapid prototyping. More importantly, the visualisation capabilities of the selected editor are explicit and supported by internal and external features (like OWLViz, OntoGraf, and the ProtégéVOWL plugin). Finally, the inferencing capabilities provided by default and add-in reasoners (like HermiT, FACT++, and Pellet reasoner) also make it suitable for application in this research [69, 71]. The selected features of the Protégé ontology environment for this research are tabulated (Table 3.2).

3.4.7 Ontology Engineering Methods

Ontologies, a critical component of the semantic web, require an engineering methodology as a guideline for ontology development. Thus far, there is no well-defined

Table 3.2 Selected Features of Protégé

Factors	*Selected features*
Language	RDF/OWL
Visualisation tool	OWLViz and OntoGraf
Reasoner type	Pellet Reasoner
Query language	SPARQL
Data importing tool	Cellfie

Source: Drafted by the author based on descriptions provided in the literature

methodology to develop an ontological model. So, ontology designer adopts their own modelling rules, principles, and design criteria depending on the ontology requirements and applications. This absence of standard guidelines has led to the slow evolution of shared ontologies and the extension and reuse of ontologies, hence reducing their applications [72]. However, some foundational ontology engineering methods (OEMs) that guide ontology development are presented in the literature. So, this section provides an overview of some of the foundational OEMs proposed in the literature (Table 3.3) and justifications for selecting a specific method. One of the older methodologies was developed by [73] for the Cyc project to create a sizeable knowledge base system. Following this method, various OEMs were developed in the following years, including the Uschold and King's, TOVE, METHONTOLOGY, simple knowledge engineering method (SKEM) or Ontology Development 101, and NeOn.

Table 3.3 shows that the processes involved in ontology modelling across different methodologies are more or less similar. The critical path adopted in the listed methods starts with specifying scenarios and objectives (determine the domain, aim, and scope of the ontology), followed by conceptualisation, formalisation, implementation, and maintenance. Of all methods, the ontology modelling in this research is guided by the SKEM methodology. Figure 3.2 depicts the steps associated with SKEM methodology and corresponding OEM stages. The critical reasons for selecting SKEM, among other methods, include the ease of understanding and use, accounting for reusing existing ontological resources, high level of specifications, flexibility, iterative and agile nature, quality documentation about the methodological process, and its integration with Protégé ontology editor (which is the selected tool for ontology modelling). Besides, the literature emphasises that this methodology is prevalent and broadly incorporated into various domains [80, 81] and has been highly cited to date [82].

The critical aspect is that ontology development is not linear but requires iteration at each modelling stage until it satisfies the user's needs. So, no single method can be considered absolute across all ontology development processes; the selection relies highly on the ontology needs and applications [54]. So, this research accounts for the suggested steps within the SKEM as guidelines for developing the semantic CW data model.

Table 3.3 Foundational OEMs

Designer/ Developer	*Methodology*	*Year*	*Ontology development steps*
[73]	Cyc	1990	Manual codification of knowledge à Knowledge codification aided by mechanised tools à Codification enabled by mechanised tools
[74]	Uschold and King's method	1995	Identify the purpose of the ontology à Ontology building (knowledge capturing, coding, and integration) à Ontology evaluation à Documentation
[75]	Grüninger and Fox's (The TOVE – Toronto Virtual Enterprise) Modelling project	1995	Identify motivating ontology scenarios (applications)/requirements à Set informal competency questions (CQs) à Determine scope à Define terminology à Formal CQs à Define critical concepts/classes, properties, and axioms of formal ontology à Develop theories
[76]	The Kactus approach	1996	Specification of the scenario/application à Preliminary ontology design (using top-level ontologies) à Ontology Evaluation à Structuring
[72]	METHONTOLOGY	1996	Specification à Knowledge acquisition à Conceptualization à Integration à Formalization à Implementation à Evaluation à Maintenance à Documentation à Configuration management
[77]	The SENSUS	1997	Identify critical terms à Add terms to the ontology à Search for missing terms à Add missing terms manually à Structuring à Ontology building
[78]	On-to-Knowledge	2001	Feasibility study à Ontology Kick-off (Requirement specification) à Refinement (Conceptualisation and Formalisation) à Evaluation à Maintenance
[54]	SKEM or Ontology Development 101	2000	Determine the domain and scope of the ontology à Consider reusing existing ontologies à Enumerate important terms in the ontology à Define the classes and the class hierarchy à Define the properties of classes – slots à Define the facets of the slots à Create instances
[79]	The NeOn Methodology	2009	No rigid methodology, but guidelines for nine different scenarios were presented. The selected guideline is for the following scenario, 're-engineering non-ontological resources'. The steps include data gathering à Conceptual abstraction à Information exploration à Search for a suitable pattern for re-engineering non-ontological resources à Transformation à Formalisation à Implementation

Source: Drafted by the author based on descriptions provided in each ontology engineering method

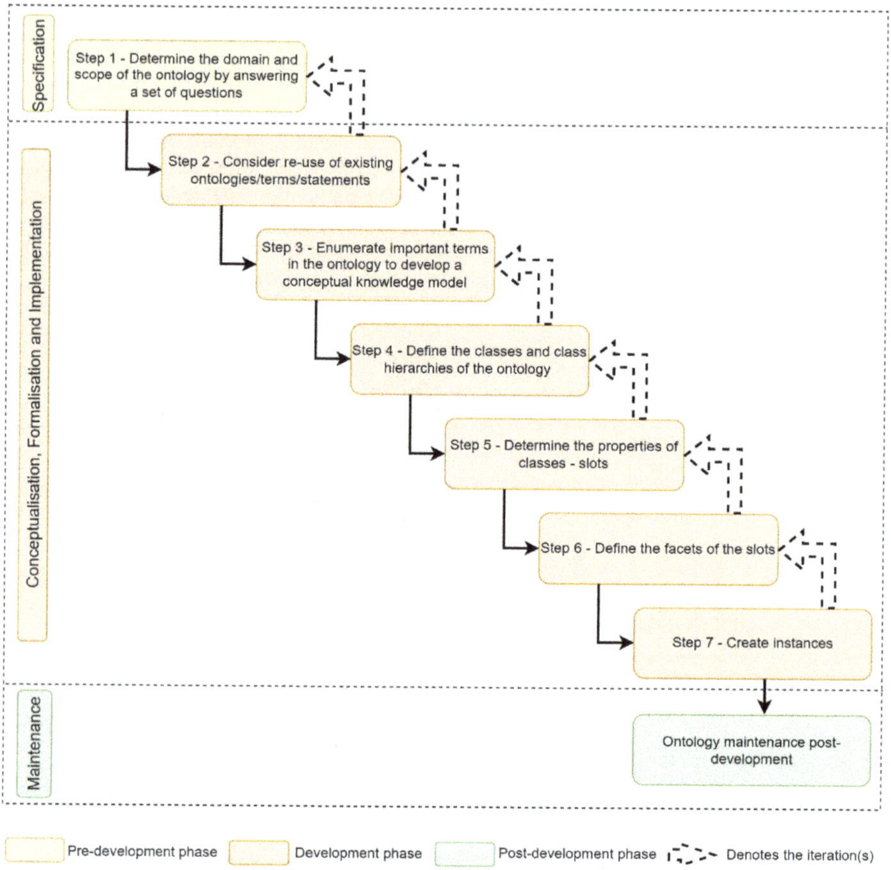

Figure 3.2 SKEM Ontology Development Steps

Source: Drafted by the author based on descriptions provided in the literature

3.4.8 Integration of BIM and SWT for Sustainable Built Environment

BIM is a process that makes a complete paradigm shift in the project workflow and delivery methods. BIM facilitates the built environment by enhancing integrated project delivery methods, collaboration, and communication across the supply chain [1]. The interoperability with data and information exchanged among different systems and participants is vital to enhance communication and collaboration across various digital platforms. In the BIM context, the interoperability is achieved using IFC. IFC can be encoded in various formats, including STEP Physical Format, XML, JavaScript Object Notation, Turtle, and RDF. The Turtle and RDF formats are developed to support the semantic web and linked data applications of IFC models using OWL. The data model type is called ifcOWL [23].

The applications of SWT in the AECO context can be seen across the whole life cycle, aiming to enhance interoperability, digital collaboration, performance analysis, and information management. Several studies investigated the applications of SWT-BIM integration to achieve a sustainable built environment. For instance, a study by Sobhkhiz, Taghaddos [83] developed a material ontology to estimate the embodied carbon emissions of materials during the product manufacturing stages and the impact of material supplier location on overall carbon levels. The LCA data was integrated with BIM to develop the semantic knowledge base and support carbon estimation. The main goal of the study was to showcase the importance of SWT in BIM-LCA integration. Another study by [84] utilised the SWT to integrate the carbon data from environmental product declaration (EPD) certificates with BIM. This study provided an overview of estimating carbon emissions during the product manufacturing stages based on the EPD data stored in the ontology. However, the carbon data from the ontology was manually imported into the BIM model as a custom parameter. Besides, the SWT applications can be seen integrated with BIM to develop a decision-support system that facilitates the development of sustainability practices from early life cycle stages [85].

Besides, the BIM-SWT integration was utilised to develop a cloud-based environment that provides real-time data to designers to assess the green building rating scores. The quality and quantity of design information in the BIM model and the semantic definitions established by SWT facilitated the assessment and scoring process [86]. Taking advantage of BIM and SWT, researchers developed a GBEOntology to overcome the gaps associated with fragmented building information and facilitate the green building evaluation process [87]. Other studies with SWT-BIM integration support energy analysis [88], project cost estimation [89], and incorporation of product manufacturer data into BIM [90] to develop an interoperable knowledge base.

3.5 SWT Functionalities Supporting CWQ

Another theoretical concept supporting the research is the SWT technologies, a promising solution to enhance data integration and collaboration through linking data. This section discusses the functionalities of SWT that support the development of the CWQ framework. The functionalities include (i) standardised data representation and exchange enhancing communication and interoperability, (ii) logical semantic inference that supports the user to draw meaningful insights, and (iii) the seamless integration of semantic web data with BIM to support an integrated and interactive CW quantification environment.

3.5.1 Standardised Data Representation and Exchange

Despite integrating several advanced ICTs in CWQ, the representation of CW data is still non-standardised and lacks a uniform structure (discussed in Section 2.7.1). The lack of standardisation also minimises the scalability of existing digital CW frameworks [91]. The non-standardised data from the CWQ frameworks significantly leads to the development of multiple data silos and comprise interoperability and information sharing. The SWT and linked data principles are widely implemented in the built environment to propose a standard data structure with a more meaningful and connected representation of information and enable data linking between domains [92]. The SWT allows the data to be connected through URIs, thus forming structured graphical data models. Thus, the semantic web data becomes meaningful, understandable, shareable, and interoperable across different software systems [56]. This further increases the reusability and scalability of data models across various scenarios. Hence, the research aims to integrate the SWT functionalities to achieve a standard data structure and enhance the integration of CW information with other assessments in the built environment.

3.5.2 Logical Semantic Inference

The semantic web has led to a paradigm shift in perceiving and utilising digital information. At the core of this revolutionary shift lies the concept of logical inference, a core functionality that allows SWT to draw conclusions based on the existing data. Relying on the two main components of SWT, ontology and rules, the inferencing process creates new data to produce new knowledge and insights [56]. An ontology is a structured data model that formally describes knowledge as a set of concepts using semantic languages like RDF and OWL and defines the relationships between the concepts. The rules are logical constructs established to infer new and meaningful facts from the existing data, guided by the reasoning engines.

Furthermore, data gaps, inconsistencies, and complexities can be detected and resolved using logical rules, enhancing data integration and interoperability. As a result, the implicit connections that are not explicitly stated in the data sources can be obtained [56]. These attributes have significant benefits across CWQ, as inadequate and inconsistent data is one of the significant gaps in the current workflow. Besides, the dynamic changes of BIM model information across the life cycle stages can easily be adapted by the semantic web model, providing accurate and granular results.

3.5.3 Data Integration With BIM

In addition to the standard CW data representation achieved using SWT, its integration with BIM is vital to enhance the communication between the BIM and the semantic data model and perform comprehensive assessments. By integrating the BIM's spatial and design information with SWT's linking, standardisation, and querying capabilities, the users would derive richer data-driven insights about CW generation levels, WM routes, and material alternatives that support waste reduction and design optimisation. Besides, the lack of interaction between CW and BIM platforms is a predominant gap, so SWT will be a robust approach to enhance communication between the two digital platforms. It further promises data transparency, automation of the process, and collaboration across the supply chain. Besides, the standardised data formats of SWT, like ifcOWL, allow the representation of BIM data in RDF graphs [23], making BIM data functional in web-based applications and simulations.

3.6 Life Cycle Assessment

In addition to BIM and SWT, the CW assessment is guided by the principles and strategies of the LCA framework. This section discusses the LCA phases and framework and the LCA attributes that aid the research in conducting a more granular and comprehensive CW analysis.

3.6.1 Phases of LCA

LCA is a systematic method applied for the quantitative assessment of environmental impacts associated with inputs and outputs of a product/process/system across its life cycle. Conducting LCA involves four main phases: goal and scope definition, life cycle inventory (LCI), life cycle impact assessment (LCIA), and interpretation [93, 94].

Goal and Scope Definition: This phase establishes the purpose, scope, and system boundaries of the LCA. It allows the assessor to identify the intended use, targeted audience, and type of product/process/system accounted for in the assessment. Accounting for these factors, the goal and scope are defined in the current research (discussed in Section 6.3.3).

Life Cycle Inventory: This phase allows the assessor to create a comprehensive, data-driven inventory of input and output flows across the entire life cycle, which can be used for environmental impact analysis in the LCIA phase. The inputs (materials, energy, and water) and outputs (emissions and waste) that will be quantified are compiled at this stage. The data source(s) for each input type is also decided in the LCI phase. The generic LCA databases and EPD certificates are widely used LCIs to extract the environmental details of manufactured products. The selection of data sources depends on the purpose of the LCA and at what project stage the LCA is conducted. The data source decides the quality, accuracy, relevance, and granularity of LCIs, which will further determine the application of the LCA results across various scenarios.

Life Cycle Impact Assessment: A critical phase in LCA, which involves three crucial tasks: (i) selecting indicators that will be assessed, (ii) assigning LCIs to each indicator, and (iii) quantifying the associated life cycle impacts. The characterised life cycle impact indicators are used to report the measured impacts. The choice of impact categories depends on environmental assessment methods like the Centre of Environmental Science Leiden-Methane (CML), Tool for the Reduction and Assessment of Chemical and Other Environmental Impact (TRACI), and Product Environmental Footprint (PEF). The first and the third methods support the formation of standard EN15804+A1 and EN15804+A2. Some of the standard impact categories emphasised across these two standards are GWP (climate change), ozone depletion, acidification potential, eutrophication, photochemical ozone creation, abiotic depletion, and water use [95].

Interpretation: According to ISO 14044:2006, three main steps are associated with this phase. It includes identifying and highlighting the significant issues faced in the LCI and LCIA phases, evaluating the completeness, sensitivity, and consistency of the results with defined goals and scope, and making recommendations based on the obtained results [94]. The insights derived in this phase will help decision-makers identify hotspots and compare alternatives, potentially guiding them to improve product selection, design, construction, use, and deconstruction. In terms of CW, this phase synthesises findings to identify opportunities for material efficiency and waste reduction. Furthermore, this stage would also guide cost-benefit analyses of various materials and WM routes.

3.6.2 LCA Framework

The LCA framework is structured to evaluate the environmental impacts across the life-cycle stages. BS EN 15978:2011 classifies life cycle stages as product manufacturing, construction/installation, in-use, and EoL stages [96]. The LCA is adopted in the built environment to analyse the impacts of materials and activities across the project life cycle. Besides, the LCA is integrated with digital technologies like BIM to overcome the information gaps in the early design stages [97, 98]. Among several life cycle impact categories, the impact of products on climate change measured using carbon metrics is widely discussed in the literature. The outputs of carbon emissions are characterised as upfront, embodied, and whole-life carbon based on the life cycle stages included in the assessment. The upfront carbon refers to greenhouse gas emissions generated during the product manufacturing and construction/installation stages. The embodied carbon refers to emissions generated across product manufacturing, in-use and EoL stages, ignoring the operational emissions. The whole life-cycle accounts for embodied carbon and the operational emissions associated with the built asset [99]. The life cycle stages are further subdivided into modules for more granular assessment. Another characterisation in LCA based on categories is also used in some cases. However, accounting for the objectives proposed in the research, stage-wise classification is adopted for reporting CWQ (Table 3.4).

3.7 LCA Attributes Aiding CWQ

LCA, a systematic environmental impact assessment approach, is accounted for in this research to develop a holistic whole life-cycle CWQ framework. The attributes and functionalities of LCA that support CWQ are: (i) the focus of LCA on the whole life cycle that supports transparent reporting, (ii) comprehensive environmental impact indicators to enhance granularity in CW results, (iii) data drive decision-support system that aids in informed decision-making, and (iv) the EoL consideration in LCA to select less impactful WM route fosters CE adoption.

3.7.1 Focus on the Whole Life Cycle

Although literature discusses the importance of CWQ for effective WM across the whole life cycle, the main shortcoming of the methodologies adopted in the existing frameworks is that they support quantification for a specific life stage(s) (discussed in Section 2.7.6). This leads to a scenario where focus on one specific

Table 3.4 Stage-Wise Classification Adopted for CWQ

Description	*LCA framework*	
	Modules	*The stage-wise classification adopted for CWQ*
Raw Material Extraction	A1	Product manufacturing
Transportation	A2	
Manufacturing	A3	
Transportation	A4	Construction/Installation
Construction and Installation	A5	
Use	B1	In use
Maintenance	B2	
Repair	B3	
Replacement	B4	
Refurbishment	B5	
Operational Energy	B6	Operational emissions
Operational Water	B7	
Deconstruction and Demolition	C1	EoL
Transport	C2	
Waste Processing	C3	
Disposal	C4	
Beyond EoL	D	Benefits and loads beyond the system boundary

Source: Drafted by the author

life cycle stage is dominated, and other stages gain less or no attention. Hence, to overcome the inconsistencies, a holistic and systematic approach emphasised by LCA to prioritise all life cycle stages is integrated into this research. This holistic scope allows the assessor to measure CW impacts and inform optimal material choices by accounting for the whole life cycle of the material. Besides, focusing on the whole life cycle ensures that the impacts are not shifted from one stage to another, thus enabling a more transparent and thorough assessment.

3.7.2 Comprehensive Environmental Impact Indicators

Witnessing the environmental impacts caused by excessive waste generation, the importance of WM is significantly increasing in the built environment. The existing CWQ approaches facilitate measuring the quantity of waste generated using weight- and volume-based estimations. Although these approaches widely support resource allocation and disposal, there is a lack of data on what impacts the generated waste is expected to create (discussed in Section 2.7.4). In such cases, the impact of lightweight materials is often overlooked despite their high environmental impact. Besides, the recent study by DEFRA emphasised that the WM should enable a shift from weight- and volume-based estimations to an impact-based waste metric system to implement sustainable solutions [100]. Hence, this research integrated the environmental indicators proposed by LCA into CWQ to allow users to explore the impacts of CW generated across the life cycle for each material type. The comprehensive set of impact categories in LCA allows users to compare each indicator and select optimal materials based on the project requirements. Furthermore, it also allows the project team to compare and benchmark the WM performance across various projects and schedules.

3.7.3 Data-Driven Decision-Support System

One of the predominant challenges in conducting CWQ is the lack of accurate and complete waste data for the entire life cycle of the material. The CW data is primarily recorded in the construction and demolition sites for management and reporting purposes at the inter-organisation level. So, there is a limited data source that is applicable to quantify the CW generation. Furthermore, existing CW data used in the literature are generic, non-standardised, and subjective, hindering their use in accurate quantification and classification [32]. Hence, the CW data from LCI is integrated with on-site waste-data collection reports to facilitate the quantification across the life cycle stages. The relevance and objectiveness of data in the manufacturer-specific EPDs promise accuracy, granularity, and transparency in the quantification results. Furthermore, the outputs obtained from this integrated and granular approach ease communication with various stakeholders across the project team.

3.7.4 Fostering CE Adoption

The CE principle emphasises the three main dimensions to enhance sustainability: resource efficiency, waste reduction, and continuous reuse of materials [101]. The LCA aims to support CE implementation by increasing the comprehensiveness of CE assessment and evaluating the environmental impacts of various CE strategies across the material life cycle stages [102]. Similarly, integrating EoL stages in the LCA-based decision-making encourages the assessor to consider the CW impacts with its EoL stages, thus promoting the reusing and recycling of materials at their higher values. Counting upon the EoL impacts, the LCA-based CWQ results would encourage designers to prioritise durability, flexibility, and deconstruction in their designs, potentially increasing the service life of buildings and reducing waste. Accounting for the fact that a specific WM route will not be an appropriate solution for all material types, the LCA-based CWQ approach would provide a comprehensive and granular set of data required to evaluate the impact of CW reuse, recycling, and disposal to make informed decisions. This further enhances how the CW strategies are developed and optimises the choices to maximise sustainability.

3.8 Summary

The two main technologies that support the framework development are BIM and SWT. In addition to these technologies, the systematic environmental impact assessment method emphasised in LCA is adopted in the research. The highlights from the chapter are as follows.

BIM:

- The LOMD 3 (LOD 200) is selected for the research.
- The output achieved from the CWQ can be attributed to 6D BIM, enabling the integration of CW data into the BIM model.
- The five main BIM attributes that support CWQ are identified. It includes life-cycle information management, automated quantity estimation, parametric modelling, interoperability with external applications, and a collaborative environment.

SWT:

- The ontology (RDF) is proposed to enable the integration of CW data from distributed data sources.
- The RDF/XML exchange format was selected for the research because of its computability across all OWL2 tools.
- The SKEM is adopted for ontology development.
- Protégé ontology environment is used to model the ontology.
- SPARQL is used to query the knowledge from the ontology.
- The main attributes of SWT supporting the research include (i) standardised data representation and exchange, (ii) logical semantic inference, and (iii) data integration with BIM.

LCA

- The four phases proposed in LCA are adopted in the research to ensure comprehensive and transparent reporting.
- Stage-wise LCA categorisation is adopted for CW quantification.
- The main attributes of LCA supporting the framework development include (i) the focus of LCA on the whole life cycle, (ii) comprehensive environmental impact indicators, (iii) data drive decision-support system, and (iv) fostering CE adoption.
- The research's core principles are interoperability, automation, and comprehensiveness, which will be achieved through the integration of BIM, SWT, and LCA frameworks.

References

1. C.M. Eastman, P. Teicholz, R. Sacks and K. Liston, *BIM handbook: A guide to building information modeling for owners, managers, designers, engineers and contractors*, 2011. John Wiley & Sons. ISBN: 978-1-119-28753-7.
2. ISO, *Organization and digitization of information about buildings and civil engineering works, including building information modelling (BIM). Information management using building information modelling – Concepts and principles*, 2019.

https://knowledge.bsigroup.com/products/organization-and-digitization-of-information-about-buildings-and-civil-engineering-works-including-building-information-modelling-bim-information-management-using-building-information-modelling-concepts-and-principles/standard. Last access: 28 June 2023.
3. B. Hardin and D. McCool, *BIM and construction management: Proven tools, methods, and workflows*, 2015. John Wiley & Sons. ISBN: 978-1-118-94276-5.
4. M. Ibrahim, R. Krawczyk and G. Schipporeit, Two approaches to BIM: A comparative study. *Architecture in the Network Society*, 2004: pp. 610–616.
5. J.D.C. Thomsen, D. Dunne, W. Lichtig, *Managing integrated project delivery*, 2009. Construction Management Association of America (CMAA), McLean, VA, p. 105. https://lean-construction-gcs.storage.googleapis.com/wp-content/uploads/2022/08/08161051/CMAA_Managing_Integrated_Project_Delivery_1–1.pdf. Last access: 02 December 2023.
6. D.C. Kent and B. Becerik-Gerber, Understanding construction industry experience and attitudes toward integrated project delivery. *Journal of Construction Engineering and Management*, 2010. 136(8): pp. 815–825. https://doi.org/10.1061/(ASCE)CO.1943–7862.0000188
7. P. Raisbeck, R. Millie and A. Maher, Assessing integrated project delivery: A comparative analysis of IPD and alliance contracting procurement routes. *Management*, 2010. 1019: p. 1028.
8. S. Azhar, Building information modeling (BIM): Trends, benefits, risks, and challenges for the AEC industry. *Leadership and Management in Engineering*, 2011. 11(3): pp. 241–252. https://doi.org/10.1061/(ASCE)LM.1943–5630.0000127
9. O. Disney, M. Roupé, M. Johansson and A.D. Leto, Embracing BIM in its totality: A total BIM case study. *Smart and Sustainable Built Environment*, 2022. 13(3): pp. 512–531. https://doi.org/10.1108/SASBE-06-2022-0124
10. M. Bew and M. Richards, BIM maturity model, in *Construct IT autumn 2008 members' meeting, Brighton, UK*, 2008.
11. HM Government, *Digital built Britain, level 3 building information modelling – Strategic plan*, 2015. https://assets.publishing.service.gov.uk/government/uploads/system/uploads/attachment_data/file/410096/bis-15–155-digital-built-britain-level-3-strategy.pdf. Last access: 11 March 2023.
12. L. Ding, Y. Zhou and B. Akinci, Building information modeling (BIM) application framework: The process of expanding from 3D to computable nD. *Automation in Construction*, 2014. 46: pp. 82–93. https://doi.org/10.1016/j.autcon.2014.04.009
13. A. Koutamanis, Dimensionality in BIM: Why BIM cannot have more than four dimensions? *Automation in Construction*, 2020. 114: p. 103153. https://doi.org/10.1016/j.autcon.2020.103153
14. S. Kaewunruen, J. Sresakoolchai and Z. Zhou, Sustainability-based lifecycle management for bridge infrastructure using 6D BIM. *Sustainability*, 2020. 12(6): p. 2436. https://doi.org/10.3390/su12062436
15. F.J. Montiel-Santiago, M.J. Hermoso-Orzáez and J. Terrados-Cepeda, Sustainability and energy efficiency: BIM 6D. Study of the BIM methodology applied to hospital buildings. Value of interior lighting and daylight in energy simulation. *Sustainability*, 2020. 12(14): p. 5731. https://doi.org/10.3390/su12145731
16. P. Yung and X. Wang, A 6D CAD model for the automatic assessment of building sustainability. *International Journal of Advanced Robotic Systems*, 2014. 11(8): p. 131. https://doi.org/10.5772/58446
17. R. Charef, H. Alaka and S. Emmitt, Beyond the third dimension of BIM: A systematic review of literature and assessment of professional views. *Journal of Building Engineering*, 2018. 19: pp. 242–257. https://doi.org/10.1016/j.jobe.2018.04.028
18. NBS, *BIM dimensions – 3D, 4D, 5D, 6D BIM explained*, 2021. https://www.thenbs.com/knowledge/bim-dimensions-3d-4d-5d-6d-bim-explained. Last access: 06 October 2024.

19. C. Harrison and D. Thurnell, BIM implementation in a New Zealand consulting quantity surveying practice. *International Journal of Construction Supply Chain Management*, 2015. 5(1): pp. 1–15. https://doi.org/10.14424/ijcscm501015-01-15
20. A. Marta, B. Stefano, B. Carlo and M. Filippo, 7D BIM for sustainability assessment in design processes: A case study of design of alternatives in severe climate and heavy use conditions. *Architecture and Engineering*, 2019. 4(2): pp. 3–12. https://doi.org/10.23968/2500-0055-2019-4-2-3-12
21. buildingSMART, *Industry foundation classes (IFC) – An introduction*. https://technical.buildingsmart.org/standards/ifc. Last access: 13 March 2023.
22. buildingSMART, *IFC formats*. https://technical.buildingsmart.org/standards/ifc/ifc-formats/. Last access: 13 March 2023.
23. P. Pauwels and W. Terkaj, EXPRESS to OWL for construction industry: Towards a recommendable and usable ifcOWL ontology. *Automation in Construction*, 2016. 63: pp. 100–133. https://doi.org/10.1016/j.autcon.2015.12.003
24. P. Teicholz, *BIM for facility managers*, 2013. John Wiley & Sons. ISBN: 1118417623.
25. S. Lavy and S. Jawadekar, A case study of using BIM and COBie for facility management, in *International journal of facility management*, 2014. pp. 13–27. https://d1wqtxts1xzle7.cloudfront.net/37943562/A_Case_Study_of_Using_BIM_and_COBie_for_Facility-libre.pdf?1434703636=&response-content-disposition=inline%3B+filename%3DA_Case_Study_of_Using_BIM_and_COBie_for.pdf&Expires=1733765794&Signature=BnWiYYZ0DTkDyHWR6Xl~EduwEO8hgxeN-YHMXXzX-oa0rXnDQzYi1fsYjamoWHvenn-2FYAc8BF00L7AeK8wPSqGWK8Xrz6Mqbtja~7i~4wptyEkvf~hiTkCYPI7-Mcc0IHz1Ft3D-~SifmE5i14LO1hHx1OXcGdhTJ3~a6dk-6Ld0QJLzXzdq9tUN~llHrVuFfdiEdFXmZSyY9RVeupW6mIvRWcq59LV~meaMbLcxhyTBvJZKclrh4vaabMEU5i~6ZFYvXNl9TgdUO-A-JC6gN66ciybvYjwWWSHVbR9c90AWfiEt0UoZYhKczltD-sTpKGxQt7T2WJlocadmhqcg__&Key-Pair-Id=APKAJLOHF5GGSLRBV4ZA. Last access: 08 December 2024.
26. B. East, Using COBie, in *BIM for facility managers*, 2013: pp. 107–143. https://doi.org/10.1002/9781119572633.ch5
27. US General Services Administration, *GSA BIM guide 05 – Energy performance*, 2015. https://www.gsa.gov/real-estate/design-and-construction/3d4d-building-information-modeling/bim-guides/bim-guide-05-energy-performance. Last access: 13 March 2023.
28. E. Gonzalez, J.D. Piñeiro, J. Toledo, R. Arnay and L. Acosta, An approach based on the ifcOWL ontology to support indoor navigation. *Egyptian Informatics Journal*, 2021. 22(1): pp. 1–13. https://doi.org/10.1016/j.eij.2020.02.008
29. NBS, *10th annual BIM report 2020*, 2020. https://www.thenbs.com/PublicationIndex/documents/details?Pub=RIBAE&DocID=329337. Last access: 13 March 2023.
30. G. Lee, R. Sacks and C.M. Eastman, Specifying parametric building object behavior (BOB) for a building information modeling system. *Automation in Construction*, 2006. 15(6): pp. 758–776. https://doi.org/10.1016/j.autcon.2005.09.009
31. S. Sivashanmugam, S. Rodriguez, F.P. Rahimian, F. Elghaish and N. Dawood, Enhancing information standards for automated construction waste quantification and classification. *Automation in Construction*, 2023. 152: p. 104898. https://doi.org/10.1016/j.autcon.2023.104898
32. S. Sivashanmugam, S. Rodriguez Trejo and F.P. Rahimian, BIM-integrated semantic framework for construction waste quantification and optimisation. *Automation in Construction*, 2024. 168: p. 105842. https://doi.org/10.1016/j.autcon.2024.105842
33. F. Fu, *Design and analysis of tall and complex structures*, 2018. Butterworth-Heinemann. ISBN: 0081011210.
34. McGraw Hill Construction, *The business value of BIM – Getting building information modeling to the bottom line*, 2009. https://images.autodesk.com/adsk/files/final_2009_bim_smartmarket_report.pdf. Last access: 15 March 2023.

35. A. Grilo and R. Jardim-Goncalves, Value proposition on interoperability of BIM and collaborative working environments. *Automation in Construction*, 2010. 19(5): pp. 522–530. https://doi.org/10.1016/j.autcon.2009.11.003
36. H.J. Moon, M.S. Choi, S.K. Kim and S.H. Ryu, Case studies for the evaluation of interoperability between a BIM based architectural model and building performance analysis programs, in *Proceedings of 12th conference of international building performance simulation association*, 2011. https://doi.org/10.26868/25222708.2011.1510
37. O.O. Akinade and L.O. Oyedele, Integrating construction supply chains within a circular economy: An ANFIS-based waste analytics system (A-WAS). *Journal of Cleaner Production*, 2019. 229: pp. 863–873. https://doi.org/10.1016/j.jclepro.2019.04.232
38. J.C. Cheng and L.Y. Ma, A BIM-based system for demolition and renovation waste estimation and planning. *Waste Management*, 2013. 33(6): pp. 1539–1551. https://doi.org/10.1016/j.wasman.2013.01.001
39. E.W. Cheng, H. Li and P. Love, Establishment of critical success factors for construction partnering. *Journal of Management in Engineering*, 2000. 16(2): pp. 84–92. https://doi.org/10.1061/(ASCE)0742–597X(2000)16:2(84)
40. N. Gudienė, A. Banaitis, N. Banaitienė and J. Lopes, Development of a conceptual critical success factors model for construction projects: A case of Lithuania. *Procedia Engineering*, 2013. 57: pp. 392–397. https://doi.org/10.1016/j.proeng.2013.04.051
41. S. Azhar, M. Khalfan and T. Maqsood, Building information modeling (BIM): now and beyond. *Australasian Journal of Construction Economics and Building*, 2012. 12(4): pp. 15–28. https://doi.org/10.5130/ajceb.v12i4.3032
42. D. Bryde, M. Broquetas and J.M. Volm, The project benefits of building information modelling (BIM). *International Journal of Project Management*, 2013. 31(7): pp. 971–980. https://doi.org/10.1016/j.ijproman.2012.12.001
43. S. Sivashanmugam, S. Rodriguez, F.P. Rahimian and N. Dawood, Maximising the construction waste reduction potential-barriers and catalysts, in *Proceedings of the institution of civil engineers-civil engineering*, 2022. Thomas Telford Ltd. https://doi.org/10.1680/jcien.22.00163
44. T. Berners-Lee, J. Hendler and O. Lassila, The semantic web. *Scientific American*, 2001. pp. 34–43. https://www-sop.inria.fr/acacia/cours/essi2006/Scientific%20American_%20Feature%20Article_%20The%20Semantic%20Web_%20May%202001.pdf. Last access: 02 March 2023.
45. W3C, *W3C semantic web frequently asked questions*, 2009. https://www.w3.org/2001/sw/SW-FAQ. Last access: 08 December 2024.
46. N. Shadbolt, T. Berners-Lee and W. Hall, The semantic web revisited. *IEEE Intelligent Systems*, 2006. 21(3): pp. 96–101. https://doi.org/10.1109/MIS.2006.62
47. G. Antoniou and F. Van Harmelen, *A semantic web primer*, 2004. MIT Press. ISBN: 9780262012102.
48. Tim-Berners-Lee. 2006. https://www.w3.org/RDF/Metalog/docs/sw-easy. Last access: 08 December 2024.
49. J.K. Korpela, *Unicode explained*, 2006. O'Reilly Media, Inc. ISBN: 059610121X.
50. T. Berners-Lee, R. Fielding and L. Masinter, *Uniform resource identifier (URI): Generic syntax*, 2005. https://datatracker.ietf.org/doc/html/rfc3986. Last access: 02 March 2023.
51. RIF Working Group, *Rule interchange format (RIF)*, 2010. https://www.w3.org/2001/sw/wiki/RIF. Last access: 18 March 2023.
52. A. Maedche and S. Staab, Ontology learning for the semantic web. *IEEE Intelligent Systems*, 2001. 16(2): pp. 72–79. https://doi.org/10.1109/5254.920602
53. T.R. Gruber, A translation approach to portable ontology specifications. *Knowledge Acquisition*, 1993. 5(2): pp. 199–220. https://doi.org/10.1006/knac.1993.1008
54. N.F. Noy and D.L. McGuinness, *Ontology development 101: A guide to creating your first ontology*, 2001. Stanford Knowledge Systems Laboratory. https://protege.stanford.

edu/publications/ontology_development/ontology101.pdf. Last access: 08 December 2024.

55. K. Giri, Role of ontology in semantic web. *DESIDOC Journal of Library & Information Technology*, 2011. 31(2). https://doi.org/10.14429/djlit.31.2.863
56. D. Allemang and J. Hendler, *Semantic web for the working ontologist: Effective modeling in RDFS and OWL*, 2011. Elsevier. ISBN: 9780123859662.
57. D. Fensel, F. Van Harmelen, I. Horrocks, D.L. McGuinness and P.F. Patel-Schneider, OIL: An ontology infrastructure for the semantic web. *IEEE Intelligent Systems*, 2001. 16(2): pp. 38–45. https://doi.org/10.1109/5254.920598
58. M. Bilal, L.O. Oyedele, J. Qadir, K. Munir, S.O. Ajayi, O.O. Akinade, H.A. Owolabi, H.A. Alaka and M. Pasha, Big Data in the construction industry: A review of present status, opportunities, and future trends. *Advanced Engineering Informatics*, 2016. 30(3): pp. 500–521. https://doi.org/10.1016/j.aei.2016.07.001
59. F. Manola, E. Miller and B. McBride, RDF primer, in *W3C recommendation*, 2004. p. 6. https://www.w3.org/TR/rdf-primer/. Last access: 02 December 2023.
60. D.L. McGuinness and F. Van Harmelen, OWL web ontology language overview, in *W3C recommendation*, 2004. p. 2004. https://www.w3.org/TR/owl-features/. Last access: 08 December 2024.
61. S. Bechhofer, F. Van Harmelen, J. Hendler, I. Horrocks, D.L. McGuinness, P.F. Patel-Schneider and L.A. Stein, OWL web ontology language reference, in *W3C recommendation*, 2004. pp. 1–53. https://www.w3.org/TR/owl-ref/. Last access: 07 March 2023.
62. W3C, *OWL 2 web ontology language – Document overview* (Second Edition), 2009. https://www.w3.org/TR/owl2-overview/. Last access: 20 March 2023.
63. J. Bailey, F. Bry, T. Furche and S. Schaffert, Web and semantic web query languages: A survey, in *Reasoning web: First international summer school 2005, Msida, Malta, July 25–29, 2005, Revised lectures*, 2005. pp. 35–133. ISBN: 978-3-540-31675-6.
64. P. Haase, J. Broekstra, A. Eberhart and R. Volz, A comparison of RDF query languages, in *ISWC*, 2004.Springer. https://doi.org/10.1007/978-3-540-30475-3_35
65. M.J. O'Connor and A.K. Das, *SQWRL: A query language for OWL*, 2009. OWLED.
66. B. DuCharme, *Learning SPARQL: querying and updating with SPARQL 1.1*, 2013. O'Reilly Media, Inc. ISBN: 978–1449371432.
67. W3C, *Semantic web development tools*, 2017. https://www.w3.org/2001/sw/wiki/Tools. Last access: 20 March 2023.
68. O. Corcho, M. Fernández-López and A. Gómez-Pérez, Methodologies, tools and languages for building ontologies. Where is their meeting point? *Data & Knowledge Engineering*, 2003. 46(1): pp. 41–64. https://doi.org/10.1016/S0169–023X(02)00195–7
69. B. Kapoor and S. Sharma, A comparative study ontology building tools for semantic web applications. *International Journal of Web & Semantic Technology*, 2010. 1(3): pp. 1–13. https://doi.org/10.5121/ijwest.2010.1301
70. R. Mizoguchi and K. Kozaki, Ontology engineering environments, in *Handbook on ontologies*, 2009. Springer. pp. 315–336. ISBN: 978-3-540-24750-0.
71. E. Alatrish, Comparison some of ontology editors, in *Journal of management information systems*, 2013. pp. 018–024. https://ef.uns.ac.rs/mis/archive-pdf/2013%20-%20No2/MIS2013-2-4.pdf. Last access: 08 December 2024.
72. M. Fernández-López, A. Gómez-Pérez and N. Juristo, *Methontology: From ontological art towards ontological engineering*, 1997. https://cdn.aaai.org/Symposia/Spring/1997/SS-97-06/SS97-06-005.pdf. Last access: 10 September 2023.
73. D.B. Lenat and R.V. Guha, *Building large knowledge-based systems; representation and inference in the Cyc project*, 1989. Addison-Wesley Longman Publishing Co., Inc. ISBN: 0201517523.
74. M. Uschold and M. King, *Towards a methodology for building ontologies*, 1995. Citeseer. https://citeseerx.ist.psu.edu/document?repid=rep1&type=pdf&doi=98304f357fb8e75aa37e5b754e905dcb94570202. Last access: 08 December 2024.

75. M. Gruninger, Methodology for the design and evaluation of ontologies, in *Proc. IJCAI'95, workshop on basic ontological issues in knowledge sharing*, 1995. https://www.researchgate.net/publication/2288533_Methodology_for_the_Design_and_Evaluation_of_Ontologies. Last access: 15 December 2024.
76. A. Bernaras, Building and reusing ontologies for electrical network applications. *Proc. of ECAI 96, 1996*, 1996. https://www.tib.eu/en/search/id/BLCP%3ACN015300062/Building-and-Reusing-Ontologies-for-Electrical/
77. B. Swartout, R. Patil, K. Knight and T. Russ, Toward distributed use of large-scale ontologies, in *Proc. of the tenth workshop on knowledge acquisition for knowledge-based systems*, 1996. https://cdn.aaai.org/Symposia/Spring/1997/SS-97-06/SS97-06-018.pdf
78. S. Staab, R. Studer, H.-P. Schnurr and Y. Sure, Knowledge processes and ontologies. *IEEE Intelligent Systems*, 2001. 16(1): pp. 26–34. https://citeseerx.ist.psu.edu/document?repid=rep1&type=pdf&doi=d05ab46e3650c34667c7d7db30b6917f623ae706
79. M.C. Suárez-Figueroa, A. Gómez-Pérez and M. Fernández-López, The NeOn methodology for ontology engineering, in *Ontology engineering in a networked world*. 2011, Springer. pp. 9–34. ISBN: 978-3-642-24794-1.
80. T. Elsaleh, S. Enshaeifar, R. Rezvani, S.T. Acton, V. Janeiko and M. Bermudez-Edo, IoT-Stream: A lightweight ontology for internet of things data streams and its use with data analytics and event detection services. *Sensors*, 2020. 20(4): p. 953. https://doi.org/10.3390/s20040953
81. A. Rhayem, M.B.A. Mhiri and F. Gargouri, Semantic web technologies for the internet of things: Systematic literature review. *Internet of Things*, 2020. 11: p. 100206. https://doi.org/10.1016/j.iot.2020.100206
82. A. Stadnicki, F.F. Pietroń and P. Burek, Towards a modern ontology development environment. *Procedia Computer Science*, 2020. 176: pp. 753–762. https://doi.org/10.1016/j.procs.2020.09.070
83. S. Sobhkhiz, H. Taghaddos, M. Rezvani and A.M. Ramezanianpour, Utilization of semantic web technologies to improve BIM-LCA applications. *Automation in Construction*, 2021. 130: p. 103842. https://doi.org/10.1016/j.autcon.2021.103842
84. Y. Schwartz, S. Eleftheriadis, R. Raslan and D. Mumovic, Semantically enriched BIM life cycle assessment to enhance buildings' environmental performance, in *Proceedings of the CIBSE technical symposium, Edinburgh, UK*, 2016. https://www.researchgate.net/profile/Stathis-Eleftheriadis/publication/296949281_Semantically_Enriched_BIM_Life_Cycle_Assessment_to_Enhance_Buildings'_Environmental_Performance/links/571ddf2208ae7f552a4a7dd7/Semantically-Enriched-BIM-Life-Cycle-Assessment-to-Enhance-Buildings-Environmental-Performance.pdf. Last access: 08 December 2023.
85. E. Petrova, P. Pauwels, K. Svidt and R.L. Jensen, Towards data-driven sustainable design: Decision support based on knowledge discovery in disparate building data. *Architectural Engineering and Design Management*, 2019. 15(5): pp. 334–356. https://doi.org/10.1080/17452007.2018.1530092
86. D. Zhang, J. Zhang, J. Guo and H. Xiong, A semantic and social approach for real-time green building rating in BIM-based design. *Sustainability*, 2019. 11(14): p. 3973. https://doi.org/10.3390/su11143973
87. S. Jiang, N. Wang and J. Wu, Combining BIM and ontology to facilitate intelligent green building evaluation. *Journal of Computing in Civil Engineering*, 2018. 32(5): pp. 04018039. https://doi.org/10.1061/(ASCE)CP.1943–5487.0000786
88. M. Niknam and S. Karshenas, Sustainable design of buildings using semantic BIM and semantic web services. *Procedia Engineering*, 2015. 118: pp. 909–917. https://doi.org/10.1016/j.proeng.2015.08.530
89. M. Niknam and S. Karshenas, Integrating distributed sources of information for construction cost estimating using Semantic Web and Semantic Web Service technologies. *Automation in Construction*, 2015. 57: pp. 222–238. https://doi.org/10.1016/j.autcon.2015.04.003

90. M. Niknam, F. Jalaei and S. Karshenas, Integrating BIM and product manufacturer data using the semantic web technologies. *Journal of Information Technology in Construction*, 2019. 24: pp. 424–439. https://www.itcon.org/paper/2019/22
91. N. Rodrigo, H. Omrany, R. Chang and J. Zuo, Leveraging digital technologies for circular economy in construction industry: A way forward. *Smart and Sustainable Built Environment*, 2024. 13(1): pp. 85–116. https://doi.org/10.1108/SASBE-05-2023-0111
92. P. Pauwels, S. Zhang and Y.-C. Lee, Semantic web technologies in AEC industry: A literature overview. *Automation in Construction*, 2017. 73: pp. 145–165. https://doi.org/10.1016/j.autcon.2016.10.003
93. ISO, *Environmental management – Life cycle assessment – Principles and framework*, 2006. https://www.iso.org/obp/ui/#iso:std:iso:14040:ed-2:v1:en. Last access: 08 December 2024.
94. ISO, *Environmental management – Life cycle assessment – Requirements and guidelines*. 2006. https://www.iso.org/standard/38498.html. Last access: 29 October 2024.
95. OneClickLCA, *Impact assessment categories (CML, TRACI, and PEF)*, 2024. https://oneclicklca.zendesk.com/hc/en-us/articles/360015036980-Impact-Assessment-Categories-CML-TRACI-and-PEF. Last access: 29 October 2024.
96. British Standards Institution (BSI), *BS EN 15978:2011 – Sustainability of construction works. Assessment of environmental performance of buildings. Calculation method*, 2011. https://knowledge.bsigroup.com/products/sustainability-of-construction-works-assessment-of-environmental-performance-of-buildings-calculation-method?version=standard. Last access: 11 October 2024.
97. A. Kamari, B.M. Kotula and C.P.L. Schultz, A BIM-based LCA tool for sustainable building design during the early design stage. *Smart and Sustainable Built Environment*, 2022. 11(2): pp. 217–244. https://doi.org/10.1108/SASBE-09-2021-0157
98. M. Röck, A. Hollberg, G. Habert and A. Passer, LCA and BIM: Integrated assessment and visualization of building elements' embodied impacts for design guidance in early stages. *Procedia Cirp*, 2018. 69: pp. 218–223. https://doi.org/10.1016/j.procir.2017.11.087
99. RICS, *Whole life carbon assessment for the built environment*, 2023. https://www.rics.org/profession-standards/rics-standards-and-guidance/sector-standards/construction-standards/whole-life-carbon-assessment. Last access: 08 December 2024.
100. Defra, *Our waste, our resources: A strategy for England*, 2018. https://assets.publishing.service.gov.uk/government/uploads/system/uploads/attachment_data/file/765914/resources-waste-strategy-dec-2018.pdf. Last access: 02 April 2023.
101. E. MacArthur, *Towards the circular economy*, 2013. pp. 23–44. https://www.ellenmacarthurfoundation.org/towards-the-circular-economy-vol-1-an-economic-and-business-rationale-for-an. Last access: 07 December 2024.
102. H. Lei, L. Li, W. Yang, Y. Bian and C.-Q. Li, An analytical review on application of life cycle assessment in circular economy for built environment. *Journal of Building Engineering*, 2021. 44: p. 103374. https://doi.org/10.1016/j.jobe.2021.103374

4 BIM-Based Semantic Construction Waste Quantification Framework

4.1 Introduction

The importance of quantification in enabling an efficient CWM is realised in the literature, but existing CWQ models still lack various limitations with data structure, data exchange, and dynamics with BIM. A BIM-integrated semantic framework is proposed to overcome the limitations of existing CWQ models. The framework ensures the CW data structure is unified, standardised, and consistent to facilitate management and reporting across the life cycle stages. It also aims to enhance the interoperability, automation, and compliance of CWQ models with BIM platforms. In addition to the data collected from the literature review, exploring the current and future scenarios of CW assessment, management, and reporting is vital. So, the industry expectations and information requirements to overcome the limitations of the existing CWQ process are also gathered from the built environment stakeholders through semi-structured interviews. The outcomes of the literature review and interviews are integrated to propose the framework, representing the mixed method design.

4.2 Philosophical Paradigm Supporting the Research

The primary objectives of the study were to explore the applications of existing CWQ models and the application of SWT integration with BIM to develop a unified CW data model. The SLR and critical review were used to meet these objectives (Chapters 2 and 3). This approach was underpinned by the interpretivist perspective. According to the interpretive approach, reality is complex and multi-layered, and a specific phenomenon can have various interpretations. It believes that reality is predefined yet subjectively constructed from social actions [1]. Interpretivism focuses on finding real meanings that may differ from individual perspectives [2]. As such, the minimisation of CW may be aided by technological enhancements, changes in the process, and the incorporation of legal frameworks. However, the main objective behind interpreting literature lies within the role of the technological dimensions of CWQ models. The models incorporated into the analysis constantly change with other individual perspectives and social constructs, implying the interpretivist position.

Furthermore, the philosophical paradigm of realism supported the research in understanding the AECO stakeholders' perspective and proposing a semantic CW

DOI: 10.1201/9781003610762-4

framework. The two types of realism are direct (empirical) and critical realism. The former emphasises that what we observe through the senses accurately portrays the real world. By contrast, the latter distinguishes between the 'observable events' and the 'real world'. Though critical realists accept that there are objective realities, they argue that relying only on positivist reasoning is insufficient to understand reality. From the critical realist views, reality ('real') is independent, external, and not directly accessible through observations [3]. Thus, critical realists undertake further analyses to understand a social construct more deeply. This, therefore, signifies the importance of a critical realist paradigm as an appropriate philosophical stance for developing the framework.

Furthermore, this paradigm functions as a general methodological framework that is subjective rather than objective. Besides, it can accommodate a range of methods for undertaking the research [1]. It allows a more profound analysis of contents and a selection of mixed research methods that fit the subject matter. These constructs of critical realism make it the best fit for the research, which is also evidenced in the following sections.

4.3 Mixed Method Approach for Framework Development

The method adopted for the design and development of the semantic framework is discussed in this section. This covers the *Step 4* of DSR (discussed in Section 1.8). The methodological choice decides whether the study follows a quantitative, qualitative, or mixed method to collect and process data. The primary factors that distinguish the quantitative and qualitative methods are that the former deals with numbers, and the latter deals with texts [2]. However, [2, 4, 5] list other factors that distinguish the quantitative and qualitative methods. Apart from the two primary methods, there is a third type called the mixed research method. This type combines qualitative and quantitative data collection and analysis in a single research. Mixed methods are alternatives that help researchers utilise the benefits of quantitative and qualitative methods while overcoming their limitations [4]. Based on the analysis, this research falls under the mixed method approach, where qualitative and quantitative methods are combined to develop the semantic framework. Qualitative research is a type of research method that allows researchers to acquire, analyse, and interpret data based on people's opinions and experiences [6]. Similarly, the critical factors concerning CW reduction potential in the industry and implications of existing WM technologies like BIM in CWQ are acquired from the literature and AECO stakeholders. Thus, the qualitative data informed the gaps and requirements of digitalised and interoperable CWQ workflow.

The mixed method allows researchers to combine the qualitative and quantitative methods in multiple ways. The major combinations would be at the data collection stage, sample selection, and data analysis [7, 8]. In this study, the qualitative data regarding the type of input and output information required to make compelling CW decisions in the design phase is analysed quantitatively. Therefore, combining qualitative and quantitative methods at the data analysis stages represents the application of mixed methods. The integration of qualitative and

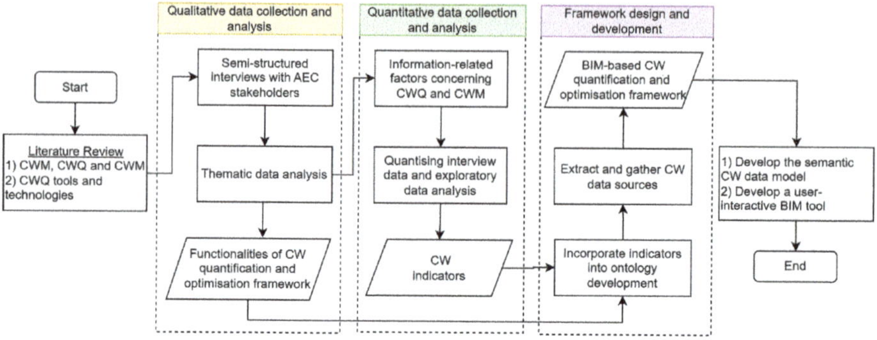

Figure 4.1 Process Flow for BIM-Based CWQ Framework

Source: Drafted by the author

quantitative approaches may differ depending on the context. As far as the study is concerned, the primary data is qualitative, collected from semi-structured interviews (Section 4.4). The limitations of existing CWQ models and expectations on future CWQ systems are measured qualitatively. In addition, the factors significant to quantifying and optimising CW are collected and analysed quantitatively. This data was collected from a quantitative response question within the semi-structured interview sessions. The responses extracted from this quantitative question were used to develop diverse CW indicators (Section 4.5). Hence, integrating one methodology into another during a single data collection process represents the concurrent embedded mixed method design.

The mixed method approach supported the research. The integration of qualitative and quantitative approaches and the process flow is depicted in Figure 4.1. The figure shows that after the literature review, the development of the solution starts with exploring the perspectives of AECO stakeholders on the current scenario of CWQ, CWM, the role of data and information in quantifying and managing CW and identifying the key functionalities of CWQ models through semi-structured interviews. This approach reflects the nature of the qualitative study. Secondly, the interview data is quantised to identify the type of information influencing CW decision-making. A granular set of CW indicators is identified and categorised using numerical ranks, reflecting the quantitative nature. Finally, the qualitative and quantitative data were analysed using thematic and exploratory analysis, respectively. The outputs of these two analyses guided the CWQ framework design and development. A detailed description of each phase is discussed in the following sections.

4.4 Exploration of Stakeholders' Perspectives on CW

The first type of primary data collected is qualitative. The critical components of the qualitative research are tabulated (Table 4.1). The data collection, sampling, and analysis techniques used to gather the qualitative data are discussed in this section.

Table 4.1 Elements of Qualitative Research

Elements of qualitative research	*Applications in the study*
Setting up the interview objective(s)	1) What are the significant barriers restraining the industry from reducing CW generation? 2) What is the role of ICTs like BIM in CW reduction? 3) What technological advancements would support the built environment in eliminating barriers associated with existing CW assessment, management, and reporting?
Data collection method	Semi-structured in-depth interviews
Sampling techniques	Purposive sampling
Data analysis	Thematic data analysis
Data presentation – Results and discussions	The factors listed in the interviews are organised as themes to highlight the significance of data, technological, management, social, and legal factors influencing CW quantification, management, and reporting across the supply chain.
Data validation	Triangulation method with secondary data gathered from the literature.

Source: Drafted by the author

4.4.1 Interview Data Collection

The qualitative data collection methods include ethnography/participant observation, interviews, and secondary sources [9, 10]. Of all methods, interviews are widely employed in qualitative research [2]. The interview types may be structured, unstructured, semi-structured, in-depth, or focus groups [2]. The approach of qualitative interviews may be objective or subjective. The former seeks only responses rather than trying to understand interviewees' perspectives, which is vice versa in the latter, where the perceptions of the participant(s) substantially influence the results [4]. Since the research objective focuses on gaining deeper insights from the stakeholders on WM, individual opinions and experiences are predicted to influence the research results. Therefore, in-depth subjective interviews are conducted individually with each participant to obtain detailed and thick descriptions of information based on their opinions and experiences.

The main approaches adopted for conducting in-depth qualitative interviews are unstructured and semi-structured [2]. By comparison, the semi-structured interview approach allows participants to express their perspectives in a deeper context without restraining their thoughts and experiences on the specific area of knowledge but with a certain degree of control [11]. It also offers flexibility in using open and close-ended questions [4]. Besides, semi-structured interview approaches are frequently adopted in exploratory studies to provide contextual background and understand the overall context of the study area, which will be helpful in the initial stages of research [4]. Since the research objective is to critically explore stakeholders' perspectives on the current waste quantification, optimisation, and management scenario, the selected interview approach is considered appropriate for this research. Thus, an in-depth semi-structured interview approach is applied to the research for its flexibility with questioning, depth of analysis, subjective reasoning, and comprehensive outcomes.

The data was collected from the stakeholders in the AECO industry, who were identified using the purposive sampling technique. In-depth interviews require a sample size ranging from 5 to 25, depending on the scope of the study [8, 12]. So, 40 stakeholders actively involved in the decision-making of CWM strategies across the building life cycle stages were invited for the interview. Among them, ten stakeholders participated in the interview sessions. The main criteria used to filter participants include their knowledge and experience about CW generation and management, BIM, recent digital innovations in the context of CWM, legal standards for CWM in the UK, and their level of involvement and decision-making in the design phases. These factors were critically analysed based on their job roles and responsibilities via a professional social networking platform. The interviews were conducted online with an average duration of 50–60 minutes for each interview. Data collected during the interviews were audio-recorded with informed consent from the participants and transcribed verbally for analysis.

The semi-structured interviews were guided by the interview guide with primary and follow-up probing questions, as suggested in the literature [11, 13]. The significant factors to account for while developing the interview guide are that the interview questions answer the research question and the themes are specific. Besides, the interviewer can incorporate follow-ups and probing questions as the interview progresses. So, there is no strict set of standard questions, but it relies on a designed theme [4, 11]. Since the research question focuses on the design phase and technological side of CWM for its significance of CW reduction, the discussions were designed to explore the following themes: significant factors affecting CW generation, existing digital technologies like BIM supporting CW causing factors in the design phase, potential benefits and limitations associated with those technologies to reduce CW generation levels, and the role of data and digitalisation in managing CW. Based on the answers received, probing discussions were included to allow the participants to provide deeper context on a specific theme supporting the research.

4.4.2 Sampling

Sampling allows researchers to reduce the amount of data needed to collect by considering the subsets/sub-groups, the representative samples of the target population [2, 4]. In this research, the stakeholders selected for interviews are samples representing the targeted population of AECO professionals in the UK. A probability (random selection) or non-probability approach (subjective selection) would facilitate the sample selection method. Non-probability sampling is categorised as convenience, snowball, purposive, and quota sampling [2, 4, 14]. The selected sampling for this research is the purposive non-probability sampling approach,

Table 4.2 Purposive Sampling Methods

Purposive sampling methods	*Description*	*Application*
Heterogenous or maximum variation sampling	Selecting individuals sharing broad and diverse characteristics relating to the research topic	For the development of themes
Homogenous sampling	Sharing similar characteristics relating to the research topic	In-depth analysis of specific case
Typical case sampling	Selecting individuals based on their likelihood of behaving like the targeted population	Illustrate typical cases
Extreme case or deviant sampling	Focus on individuals that are extreme/unusual and atypical.	Develop best-in-practice guidelines
Theoretical sampling	Selecting samples according to developing categories and emerging theories	Informs emerging concepts and theories
Expert sampling	Selecting individuals with specific expertise in the research topic	Investigating new areas and gaining knowledge of whether further research is required.

Source: Drafted by the author

which has high subjectivity and less randomness of the samples. Besides, the critical idea of purposive sampling is to focus on participants with specific characteristics to obtain rich, comprehensive information for the research [15]. The purposive sampling methods are tabulated (Table 4.2).

This research collected data from the subject matter experts active in the AECO industry but with different roles and experiences, representing heterogeneous purposive sampling characteristics. The three significant criteria used to select the participants and obtain rich quality information include their (i) knowledge and experience about digital design and construction techniques like BIM, (ii) existing innovative CWM methods and their limitations, and (iii) legal CWM practices in the UK.

4.4.3 Thematic Data Analysis

The critical part of qualitative research is to analyse and derive logical meanings from the rich yet unstructured comprehensive data collected [2, 4]. The qualitative data analysis methods include thematic analysis, content analysis, narrative analysis/inquiry, grounded theory method, template analysis, analytic induction, and discourse analysis [2, 4, 16, 17]. Among all approaches, thematic data analysis is used to analyse the transcripts of semi-structured interviews. It is a generic and flexible approach that allows researchers to emphasise 'what is said rather than how it is said' [2]. The thematic data analysis method is widely recognised for its applications across various research questions, approaches, designs, and sample sizes [16, 18].

Thematic data analysis is a structured approach for analysing, organising, describing, and presenting themes and their relationships identified within the data set [19]. A vital part of this method is that it allows the researcher to examine the perspectives of different participants, their similarities, and distinctions and to generate insightful outcomes from the data [19]. In addition, this relies on the intangible ability of the researcher to derive concepts, generate themes, and explore the relationships developing from the data [20, 21], which aligns with the research objectives. Therefore, thematic data analysis is considered more appropriate for this research to present rich quality information by establishing the perspectives of stakeholders involved in the data collection process. NVivo, a computer-aided data analysis software, is used in the research to analyse the interview data. The verbally transferred audio data is used to facilitate the thematic analysis. The six-step process followed in the thematic data analysis is guided by [19]. The steps include the following:

- **Familiarising with the data** – This is the foundation step in the thematic data analysis, which requires repeated and active reading of the complete dataset. It allows the analyst to familiarise him/herself with the overall data structure. The verbal transcripts collected from the interview are entirely read to get an overview of the dataset before coding. Each transcript was imported into the NVivo platform as separate files to support the examination.
- **Generating initial codes** – Codes are labels/units/names given to a specific number of words, sentences, and number of sentences within the transcribed data [4]. The codes are formed, constantly compared and re-read to ensure consistency within the data set. The size of the codes shows the occurrence and non-occurrence of a phenomenon.

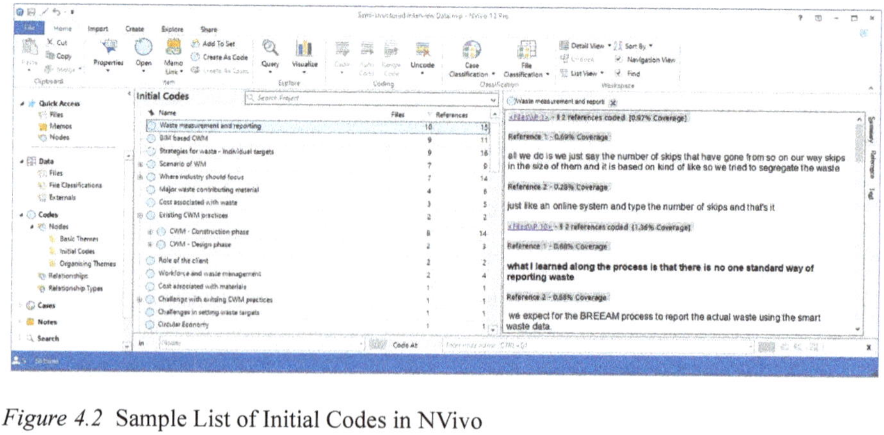

Figure 4.2 Sample List of Initial Codes in NVivo

Source: Drafted by the author

Table 4.3 Formation of Themes (Sample)

Transcripts	*Participant number*	*Codes*	*Organised themes*
'If designers can develop designs that can match with product sizes, then we can ultimately avoid offcuts, which is nearly causing 15–20% of the total waste'	P1	Standardised designs	Waste-driven designs
P10: 'We don't have reliability in the waste estimates to optimise designs, which is one of the leading challenges that need to be concentrated to integrate waste analysis into the design.'	P10	Lack of consistent and granular waste information	Digital innovations in waste estimation and management
'In terms of collaboration, convincing the client to go for reused products is challenging.'	P8	Client's resistance to secondary products	Collaboration within the supply chain

Source: Drafted by the author

- **Searching for themes** – Based on the development of initial codes, the themes are formed, aligning the appropriateness of the data. Different themes are initially created for data, information, design, procurement, workforce, client, cost, time, digital innovations, BIM, and legal aspects.
- **Reviewing and organising themes** – The themes created in the earlier steps were reviewed for clarity and organised to assist in the final analysis. Table 4.3 shows a sample of codes and organised themes from the interview data. These initial codes allow further categorisation according to the themes.
- **Mapping/Establishing relationship between themes** – The association between the themes is mapped for further interpretations (Figure 4.3).
- **Data presentation** – The organised themes and their interrelationships are presented in the following section.

4.4.4 Participant's Background

Overall, 20% of the participants were principal sustainability consultants with design and architectural backgrounds, 20% were construction site managers, 20% were sustainability managers involved in the design phase, 30% were sustainability managers and consultants involved in the construction phases, and 10% were project managers/directors. The participants had sound knowledge and experience in environmental sustainability, digital technologies, CE principles, WM practices, and legal waste policies in the UK and were actively involved in CW decision-making in the design and construction phases of building projects.

Around 60% of the participants were from large-sized international organisations with over 500 employees and a £36 million annual turnover. The remaining 40% were from organisations with construction services within the UK region.

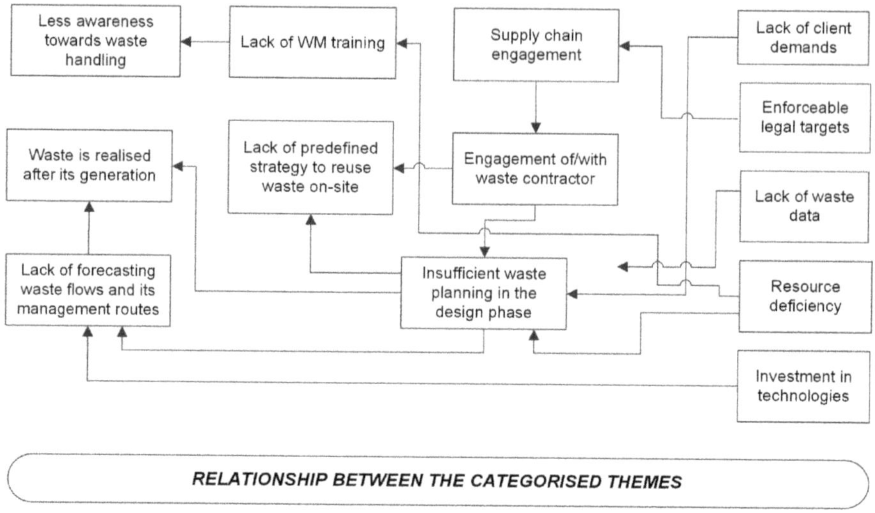

Figure 4.3 Relationship Between the Themed Categories – A Sample
Source: Drafted by the author

Meanwhile, 20% were small-sized, and the remaining 20% were medium- to large-sized construction players. Overall, the comprehensive results obtained from this population would be applied to develop robust CW quantification and management plans across the project phases.

4.4.5 Barriers and Catalysts for CW Reduction – Stakeholder Perspectives

The data analysis yielded 22 factors concerning the limitations of existing CWM methods and the potential changes required to enhance waste reduction. The factors include data, technical, management, social, and other legal changes that demand attention from the stakeholders. In addition to these factors, limitations within the current WM setting, industry requirements on prospective WM innovations, and the potential ways to meet the industry demands are discussed as four themes (Table 4.4).

4.4.5.1 Waste-Data-Driven Decisions in the Design Stages

Waste generation is influenced by design, procurement, construction, human resources, site conditions, and external factors [22]. Each factor determines the quantity and quality of waste generated. The majority of the stakeholders involved in the design and construction team observed that the design factors, including 'offcuts arising from unstandardised designs, an assortment of finishing materials, frequent design changes occurring as a result of design errors, inaccurate design detailing and overly complex designs' have an intense impact on the waste

Table 4.4 Results From Thematic Data Analysis

Themes	*Barriers of existing CWM practices*
Waste-data-driven decisions in the design stages	The CW data used to optimise designs are inaccurate and unreliable. There is a lack of digital CW information for construction products. The design-out waste principles are practically not well integrated into designs. The developed CWM plans lean towards the construction and post-construction phases. Insufficient time to plan for CWM in the design phase
Digital innovation in CW quantification and management	Lack of communication between the design, construction, and CW platforms The existing CW platforms have limited compliance with the BIM environment. Working with multiple data platforms is complex and time-consuming. Existing CW platforms require a high level of human intervention.
Integration of carbon data in the CWQ environment	Narrow vision towards carbon emissions of building materials Existing sustainable targets focus on carbon impacts. The weight-based CW estimation is primarily beneficial for measuring waste after its generation. The granularity of CW information is low.
Collaboration within the supply chain	The fragmented CWM system leads to a lack of collaboration within the project team. The CW strategies and legal CW targets are not well defined. Inadequate support from the client Resources spent on CWM are minimal. Decisions are generally made from a financial perspective. Lack of awareness about CW generation and its impacts.

Source: Drafted by the author

generation levels. It aligns with previous studies, stating that one-third of the waste-causing factors are design incurred [23, 24].

Besides, designers acknowledge the ability of BIM in the pre-construction stages to reduce waste generation. The BIM applications, including design visualisation, clash-detection features, and accurate material information, facilitate the designers and quantity surveyors to handle the design factors causing waste. Previous studies have also proven the efficiency of BIM in CW reduction [25–27]. For instance, [27] proved that an average of 10% waste reduction on-site can be achieved through clash detection and design visualisation using BIM. However, it can be inferred from the interviews that most CWM plans employed in the industry, like reuse and recycling, primarily lean towards the construction phase, and the efforts taken in the planning and design phases to reduce waste are trivial. This scenario should significantly improve to have a stabilised and reliant workflow across the project life cycle stages (Figure 4.4).

In addition, designers observe insufficient waste data as a significant barrier to achieving design optimisation and standardisation. For example, a participant (P1) mentioned, 'Typically, waste is estimated as 10%-12% of the total materials ordered, but the true values of waste quantities are still uncertain'. Besides, the participant emphasised that 'measuring waste with respect to the design, identifying the hotspots, and considering like what can we do with waste, how we can reuse it, and coming back to the Waste Hierarchy will expand our options to optimise designs'.

The stakeholders also suggest that accurate and granular CW data will pave the way for developing competent WM strategies in the construction phases. The findings align with previous studies, where measuring the waste quantities, type of waste, and time and place of generation is considered vital for enhanced CWM [24, 28, 29]. Although several efforts have been put into quantifying CW using the design information, researchers emphasise that insufficient and inaccurate waste data are critical barriers to detailed estimation [30, 31]. Based on the results from

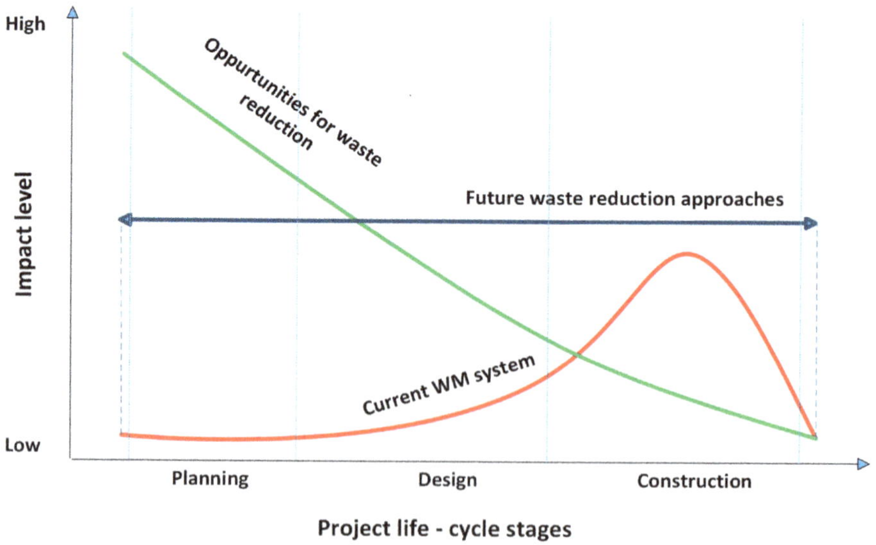

Figure 4.4 Stakeholders' Perspective on Current and Future Waste Reduction Approaches

Source: Drafted by the author

the European Commission on the data quality of EU-28 member states, the overall CW data quality of the UK is considered 'moderate' due to some ambiguities in waste reports [32]. Hence, the prospective step should be taken to build alleviated CWM measures from the initial project phases to eliminate waste generation. Nevertheless, the information required to support decision-making should also be organised to integrate CW factors into building designs.

4.4.5.2 Digital Innovation in CW Quantification and Management

Measuring waste from the developed design is significant; nevertheless, the features of CW platforms are anticipated to set a specific standard for their management. Although most participants were clear about insufficient CWM strategies in the design phase, very few highlighted the need for CW platforms to support waste quantification in the early project stages. Despite the adoption of these platforms, the significant limitations emphasised by the stakeholders include

> high level of human intervention, lack of communication between design and waste knowledge bases, applicable only in the late design stages, insufficient time to adopt necessary revisions into the design, and forecasted waste results are not aligning with actual waste outputs.

The construction industry deals with tons of heterogeneous data fragmented across the supply chain [33, 34]. The Institution of Engineering and Technology [35] identifies the eight distinct types of product data adopted across the AECO project life cycle stages, which include manufacturing, dimensional, logistics, cost, maintenance, identification, performance, and sustainability data. Past studies [33, 36] suggest that the complexity of handling numerous data can be resolved by adopting an integrated open data format to enable interaction between the design, planning, construction, and other external platforms. Besides, buildings are complex entities packed with an assorted range of materials [37]. So, the literature [38] suggests that BIM models can provide accurate and detailed information on each building component's geometric and physical properties essential for informed decision-making. Moreover, open BIM data formats like IFC facilitate the data exchange between multiple BIM platforms [39]. Furthermore, the NBS report states that 71% of construction professionals in the UK recognise BIM for various applications, ranging from 3D modelling to sustainability analysis and facilities management [40].

Hence, a digital tool that requires less human intervention and is highly interoperable with BIM platforms can expand the prospects for CW quantification and management across the life-cycle stages. This finding aligns with the studies [26, 30, 33, 41, 42], demonstrating the importance of automation and interoperability through digital technologies like BIM, machine learning, big data, and blockchain in the CW context (Figure 4.5). Therefore, portraying these enhancements in the waste platforms allows the project team to shift from a reactive to a proactive CWM.

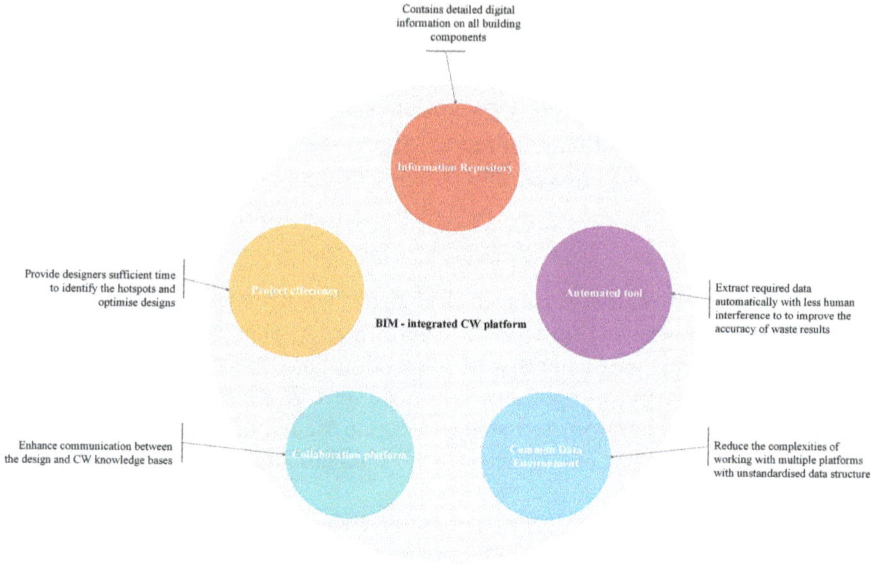

Figure 4.5 Functionalities of BIM-Integrated CW Tools

Source: Drafted by the author

4.4.5.3 Integration of Carbon Data in the CWQ Environment

WGR is a key indicator to benchmark waste performance across the project life cycle [43]. The generation rates are typically measured against GFA, project cost, and duration [44, 45]. This type of estimation is broadly used in waste reports to help contractors plan for allocating appropriate resources. However, designers stress that incorporating carbon factors into the waste analysis could enhance the consistency of information utilised for decision-making across the asset's life cycle. A participant (P5) said, 'I think it will potentially muddy the waters if carbon factors are excluded from the waste analysis'. In addition to the technical circumstances, most participants stress that the carbon and economic values of waste materials would gain momentous attention from the clients due to net zero carbon targets. A participant (P9) mentioned, 'In recent times, clients are interested in carbon emissions in addition to customary factors like time and cost. So, applying carbon metrics in waste-related discussions will be more reliable and easily reach the clients'.

The impacts of waste factors in deciding carbon levels are also broadly demonstrated in the literature [46–48]. For instance, it is estimated that 17 m^3 of CW from a timber frame house generated 109 $kgCO_2e$ emissions [49]. Thus, integrating the impact values of waste would standardise the information and support informed decision-making [50]. As stated in past studies [51–53], the characterised impact values of construction products presented in the LCI databases, EPD certificates, and material passports have an influential role in accelerating LCA. Furthermore, the idea of blending the LCA methodology with the Waste Hierarchy to minimise the environmental impacts of the CWM system is also expressed in [54].

Notwithstanding the fact that LCA data repositories are helpful for CWM, a few participants state that 'methodological choices, system boundaries, carbon targets, benchmarks and data readiness' will adversely impact evaluating end-of-life emissions. This is proved by [55], stating that end-of-life scenarios within the current LCA studies are deficient because of inconsistent data and chosen system boundaries, which should be enhanced to include the whole building life cycle. Hence, a wide range of technical and environmental indicators from appropriate sources should be collected to optimise the life cycle impacts of CW.

4.4.5.4 Collaboration Within the Whole Supply Chain

Construction is a human-driven industry involving multiple stakeholders from various disciplines. A high level of fragmentation is an apparent problem that has prevailed for decades in the construction setting [56–58]. Similarly, participants observe the lack of collaboration among the project stakeholders as a critical factor affecting CWM performance throughout the project life cycle. The significant factors observed in the interview and the corresponding stakeholders influencing those factors are tabulated (Table 4.5).

Table 4.5 Supply Chain Members and CWM Factors

CWM factors	*Responsible players*
Having a common-shared goal among the construction team members	Main and sub-contractors
Digital information on construction products	Suppliers
Integrating waste factors into the design	Designers and architects
Designer's attitude to altering the design	Designers and architects
Enforceable legal standards	Government
A clear description of secondary (recycled) products	Government
Enhancing the secondary product market value	Government
Incentives to boost WM performance	Government
Increasing the trust in recyclable products	Government
Interest in environmental attributes	The Client
Choice of contractors proficient in WM	The Client
Material efficiency and WM as part of Employer's Information Requirements	The Client
Awareness about material resources and waste	The whole supply chain
Allocate sufficient time, cost, and human resources for WM planning	Project Managers
Increased transparency among the parties	The whole supply chain
Collective risk and award – stop blaming others for mistakes	The whole supply chain
Appropriate waste strategies	Policymakers
Earlier engagement of/with waste contractors	Design and construction team members

Source: Drafted by the author

It can be inferred that the role of internal and external stakeholders, including the client, suppliers, and legal authorities, significantly impacts the quality of CW decisions. A few participants emphasised that proper collaboration and communication within the project team decide the success rate of innovations. A participant (P8) said:

> As a team, we are doing our best to divert waste from landfills, and we have also initiated some steps in terms of technologies like BIM. But without proper coordination and without enough support from the client, it will be challenging for us to reach the zero waste targets.

Innovation in the AECO industry is described as the application of new ideas, technologies, products, and policies to improve project efficiency [59]. Despite innovation being critical for sustainable construction, collaboration and communication among the supply chain members determine the productivity levels of innovations [60]. The UK Construction Strategy 2025 signifies the growing need for stakeholder collaboration to accomplish the ambitious targets [61]. Further, it also emphasised that innovative actors are vital for sustainable growth in addition to innovative concepts and technologies [62].

Besides, the participants observe that environmental elements are established within projects either through the personal interest of the decision-making communities or as a requirement to meet the legal standards, which are currently deficient in the CW context. A participant (P10) claimed, 'One of the best ways to motivate the industry in this transition period would be by having clearly defined waste strategies, enforceable legal targets, any form of incentives and solid support from the client'. Likewise, most participants stress that although the client considers CWM, it is not their top priority because of upfront cost. The statement is supported by [63, 64], where the importance of cost factors stopping clients from investing in environmental aspects is at the forefront of various limitations. Besides, trust, commitment, reliability, and behavioural aspects are asserted as enablers of collaboration in the construction supply chain [56, 65, 66]. The literature justifies that the project stakeholders/actors are critical for innovation, and they could also ease the complexities associated with sustainable visions [67]. Inclusive of all factors, a coordinated effort among the supply chain, including the client, suppliers, and policymakers, can enrich the quality of CWM.

Of all the themes discussed, accurate and reliable CW information is the leading factor in enhancing their quantification and optimisation from the early design stages. Although digital technologies like BIM seem to gain attention from the industry to minimise CW generation, standardised waste data and enhanced digital communication within the design, construction, and waste platforms significantly impede its applications. Therefore, an intense effort is crucial to ensure that the integrated and accessible waste data is supplied to the project team on time to strengthen information flow structures, decision-making and collaboration among the project stakeholders.

4.5 Identification of Indicators for CW-Driven Decision-Making

The second data type used in this research is quantitative. Since semi-structured interviews allow the choice of using mixed question types, the qualitative data collected in this research is aided by open and close-ended questions. Furthermore, the qualitative data is converted to a quantitative form through quantising to analyse the CW indicators. Therefore, this section discusses the process of quantising qualitative data collected from the interviews and the results of exploratory univariate data analysis.

4.5.1 Quantising Interview Data

The research is dominated by qualitative data reports (discussed in Section 4.3). However, the identification of stakeholders' expectations about the type of information required to make solid decisions on CW and its management routes is analysed quantitatively. Hence, the qualitative data collected from the semi-structured interviews are converted into quantitative data at the analysis stage. This process of assigning empirical values to the qualitative data to convert it into a quantitative form is termed *quantizing* [7]. The conversion of qualitative into quantitative data in mixed methods research is often carried out to put qualitative data into a format amenable to empirical analysis [68], which provides numerical values to the data conceived as non-numerical. In the quantising process, the variables are introduced to statistically categorise and analyse the interview transcripts to support the quantification process while preserving the qualitative meaning of the data. Besides, quantised qualitative data facilitate pattern recognition and present data frequencies in a form that is easy to communicate [69]. Similarly, a specific set of qualitative data collected from semi-structured interviews is converted into a quantitative form for empirical analysis. This data enhancement aids in determining the frequency of occurrences of each information type across the conducted interviews. This information acts as CW indicators within the ontology.

4.5.2 Exploratory Data Analysis

The type of data analysis depends on the type of quantitative data. The two distinct types of quantitative data are categorical (also called descriptive or nominal data) and numerical data. The categorical data are further classified as dichotomous and ranked, and numerical data are further classified as discrete and continuous [4]. In comparison, Bryman [2] uses the term 'variables' instead of data types and classifies them as interval/ratio, nominal/categorical, ordinal, and dichotomous variables. Based on the authors' descriptions of data types, the data type collected for analysis is categorical, which can be sorted as categories (also called sets) but cannot be ranked. Thus, the type of information vital for the CWQ model represents the characteristics of categorical data, where the data can only be categorised but not ranked numerically. Exploratory data analysis is encouraged in the early stages of analysis to use graphs and charts to explore and present the data [4]. Furthermore, depending on

Table 4.6 List of CW Indicators

S. No.	*Category*	*Indicators*	*Reason for its inclusion*
1.	**Technical**	Product length	To match the design size with material size, reduce offcuts, and standardise designs.
2.		Product breadth	To match the design size with material size, reduce offcuts, and standardise designs.
3.		Product thickness/height	To match the design size with material size, reduce offcuts, and standardise designs.
4.		Service life	To estimate the number of replacement cycles and applications across the building life cycle.
5.		Product density	To determine the weight of the material, which in turn provides the weight of waste.
6.	**Environmental**	Use of virgin/raw materials	To demonstrate its strength and applications across various building elements.
7.		Use of reused and recycled materials	To demonstrate its strength and applications across various building elements.
8.		Waste quantity/WGR per GFA	To benchmark the generation rates based on the construction, refurbishment, and demolition areas.
9.		Waste (material) type	To understand the reactive nature of waste type (hazardous or non-hazardous).
10.		Reusability, recyclability, and waste as energy	To select an optimised WM route based on project needs and site environment.
11.		Landfilling	To select an optimised WM route and facilitate waste reporting based on legal standards.
12.		EWC	To ensure waste reporting in the waste transfer notes aligns with the waste duty of care.
13.		Quantity of packaging	To measure the secondary waste quantities (e.g. no. of plasterboards per pallet).
14.		Packaging material type	To decide the waste treatment routes of secondary materials.
15.		Embodied carbon of waste materials	To analyse the potential of CW reduction on the Net Zero carbon targets, thereby enhancing environmental sustainability.
16.		Embodied carbon of waste treatment and transportation	To select the WM route based on its environmental impacts.
17.	**Economic**	Cost of waste materials	To highlight the contributions of waste towards project cost.
18.		Cost of waste treatment and transportation	To highlight the contributions of waste towards project cost.
19.	**Generic**	Product application(s)	To reduce material diversity (Selecting a specific material that suits multiple applications would reduce waste).
20.		Manufacturer name	To purchase products from the same manufacturer to reduce multiple transportation.
21.		Manufacturing unit/location	Purchase products from the manufacturing unit near the construction site to reduce long-distance transportation and storage.
22.		Alternatives	To explore if any alternative materials with the same features could support the design.

Source: Drafted by the author

Table 4.7 Frequency Distribution of CW Indicators

Categories	*Indicators*	*Participant* *P1*	*P2*	*P3*	*P4*	*P5*	*P6*	*P7*	*P8*	*P9*	*P10*	*Frequency (n)*	*Percentage (%)*
Technical	Product length	✓	✓		✓	✓	✓	✓	✓	✓	✓	9	7
	Product breadth	✓	✓		✓	✓	✓	✓	✓	✓	✓	9	7
	Product thickness/height	✓	✓		✓	✓	✓	✓	✓	✓	✓	9	7
	Service life	✓	✓				✓	✓	✓	✓	✓	7	6
	Product density					✓					✓	3	2
Environmental	Use of raw materials	✓		✓		✓		✓			✓	5	4
	Use of reused and recycled materials	✓		✓								7	6
	Waste quantity/WGR per GFA	✓	✓	✓		✓	✓	✓	✓	✓	✓	10	8
	Waste (material) type	✓	✓	✓	✓	✓	✓	✓	✓	✓	✓	8	6
	Reusability, recyclability, and energy as waste	✓	✓	✓	✓	✓	✓	✓	✓	✓	✓	10	8
	Landfilling	✓			✓	✓	✓		✓	✓	✓	7	6
	EWC	✓		✓		✓		✓	✓		✓	6	5
	Quantity of packaging	✓	✓				✓				✓	4	3
	Packaging material type	✓	✓				✓				✓	4	3
	Embodied carbon of waste materials	✓			✓	✓	✓		✓	✓	✓	7	6
	Embodied carbon of waste transportation and treatment	✓					✓		✓	✓	✓	5	4
Economic	Cost of waste materials	✓			✓	✓						3	2
	Cost of waste treatment and transportation	✓			✓	✓						3	2
Generic	Product application(s)	✓	✓								✓	3	2
	Manufacturer name	✓					✓	✓		✓		3	2
	Manufacturing unit/location	✓						✓		✓		2	2
	Alternatives	✓									✓	1	1
Total												125	100

Source: Drafted by the author

how the variables and their relationships need to be presented, the data analysis is classified as univariate, bivariate, multi-variate, and statistical inference testing [2]. Since the purpose of the quantitative analysis in the research is to determine the frequency of occurrences of a categorical variable(s) across the conducted interviews, an exploratory univariate analysis approach is selected.

All the participants involved in the interview listed a comprehensive and diverse set of information required for waste-informed decision-making. Table 4.6 lists indicators extracted from the data and the reasons for their inclusion. All these indicators are grouped into four major categories: technical, environmental, economic, and generic descriptions of building products.

Overall, 22 distinct types of indicators are highlighted by the participants involved in the interviews. In order to determine the level of significance of each indicator, their frequencies are calculated based on how often an indicator is used across the conducted interviews. For instance, if the indicator 'product length' is highlighted by nine participants, the number of occurrences is computed as 9. The same analysis is applied to other indicators. Table 4.7 depicts the frequency distribution of the listed information types. The frequency distribution shows the diversity of indicators and their role in making waste-informed decisions. To empirically evaluate the importance of each indicator, they are grouped into four types based on their impact levels: 'critical', 'high', 'medium', and 'low'. The indicators that have frequency number ≥ 7 to ≤ 10 is considered as critically important indicators, ≥ 5 to <7 is classified as highly important, ≥ 3 to <5 is classified as moderately important, and ≥ 0 to <3 is categorised as a low-importance indicator group. The groups and associated indicators are tabulated (Table 4.8).

Table 4.8 CW Indicators and Their Groups

Groups	*Indicators*
Critically important	Product length, breadth, and thickness/height, service life, use of reused and recycled materials, waste quantity/WGR per GFA, waste (material) type, reusability, recyclability, and waste as energy, landfilling, embodied carbon of waste materials
Highly important	Use of raw materials, EWC, embodied carbon of waste treatment and transportation
Moderately important	Product density, product application(s), product manufacturing unit, and manufacturer name
Low importance	Cost of waste materials, cost of waste treatment and transportation, quantity of packaging, packaging material type, and alternatives

Source: Drafted by the author

Each indicator plays a significant role in influencing decision-making across life-cycle stages. However, accounting for the time and data constraints, the critical, high, and modest indicators are only accounted for in the research.

4.6 BIM-Based Semantic CWQ Framework

This section presents the components of the proposed BIM-based semantic CW framework. The framework comprises three main modules: the data collection or extraction module, the data mapping and standardisation module, and the BIM-based CWQ and decision-making module (Figure 4.6). The framework aims to enhance interoperability between BIM and CWQ models, unify and standardise CW information, ease the CW quantification process, and ensure collaborative working across the project life cycle stages. Below is a brief description of each module with input(s), process(es), and associated output(s).

4.6.1 Data Collection/Extraction Module

The data collection/extraction module illustrates the data types required for quantifying CW and informing optimal material choices from the early design stages of the project. The three main data clusters include building and design details, waste factors, and technical, physical, and GWP of materials. The type of data influencing the overall process was clustered based on the outcomes of the literature review (discussed in Sections 2.6 and 2.7) and stakeholder perspectives (discussed in Sections 4.4.5 and 4.5.2). Since BIM is considered a rich data source [70, 71], the first cluster, building information, including the building's geometric and functional properties, is extracted from the BIM model. The LOMD represents the level of graphical and non-graphical information within the developed BIM model to maintain data quality and ease information exchange. Although there is no linear relationship between the LOMD and project life-cycle stages, the framework is designed to work with a minimum LOMD of 3 to ensure consistency with the CW results in the early design stages. This level equals LOD 200 within the LOD framework (discussed in Section 3.2.3).

There are multiple CW data sources, including existing site waste records, statistical reports, waste disposal records, and secondary data sources to extract the waste factor of a building material [29]. Thus far, no single waste data source has been found to have broader implications across various applications. So, the accessibility, availability, and completeness of datasets should be accounted for before making a selection [37]. Two different conditions are presented in the model to obtain the waste factor. Firstly, if the extracted data is an analysed and cleaned report, it can be directly used for quantification. This includes data from CW benchmarking and data collection platforms. In contrast, the extracted data will be raw if extracted from site waste records or direct site visits. In such cases, statistical or AI-based analysis should be conducted before integrating the extracted waste data with other data clusters. The choice of analysis will depend on the volume, variety, and velocity of the data (3Vs), as suggested by [33].

The third cluster, technical, physical, and GWP of materials, is vital to provide designers with the best-optimised materials. The primary data sources for collecting these data are EPD data repositories. These databases provide detailed information that can be used to measure the environmental impacts of building materials across their whole life cycle [72]. The granularity of the data within the EPDs makes it appropriate to measure environmental impacts precisely. Despite the benefits, inconsistency and incompleteness of LCA impact values hinder the options to compare multiple EPDs of the same material type [52]. So, a detailed investigation should be conducted to select EPDs that can be compared with other EPDs of the same material without any constraints. In case of missing EPDs, generic LCA data sources of building materials are also preferred. The data source should be selected based on the LOMD of the BIM model [52]. In general, it is preferred that if the LOMD ≥ 3, product-specific EPDs are appropriate, while generic LCA reports are best fit for lower LOMDs [73]. Besides, the data extraction methods and data types differ depending on the databases selected. Although digital EPDs ease the data extraction process, accessing the technical properties of the products is limited. So, it is also important to note if the extracted EPD data holds adequate information about the generic and technical properties of the product required for quantifying CW and informing optimal material choices [74].

4.6.2 Data Mapping and Standardisation Module

Integrating the CW model with the BIM platform is essential to demonstrate seamless communication between design, construction, and waste platforms. It is essential to verify that material, waste, and EPD data used for computational analyses between various digital platforms are standardised before mapping. The difference in terminologies, data hierarchy, and variable structures limits automation and dynamics within the CWQ process [37]. Besides, the Zero Avoidable Waste report highlights that differences in material and waste codes significantly impact interoperability [75], thus hindering the accessible communication of building and CW platforms. Although the properties of these materials are similar, standard terminologies and classification systems are vital to achieving communication and automation within the process. Therefore, the Unified Classification System (Uniclass) 2015 [76] and EWC codes [77] are proposed to ensure consistency with the design, material, and CW information. In addition, API interaction within the framework modelling emphasises using unique IDs of materials within the LCA databases.

Thus, considering the limitations like varied data structures and formats, unstandardised information, and lack of dynamics within the quantification process, an ontology aided by the SWT is proposed. The diverse data collected from CW and LCA databases is standardised to an RDF format, thus leading to the development of an ontology. The ontology modelled in the RDF format is integrated with BIM via an API or an interface supported by the selected BIM modelling platform. This integration enables automation across the data extraction, CW quantification, and material optimisation. It also aims to structure the information flow and ease waste-data exchange across the building life cycle. The standardised and consistent

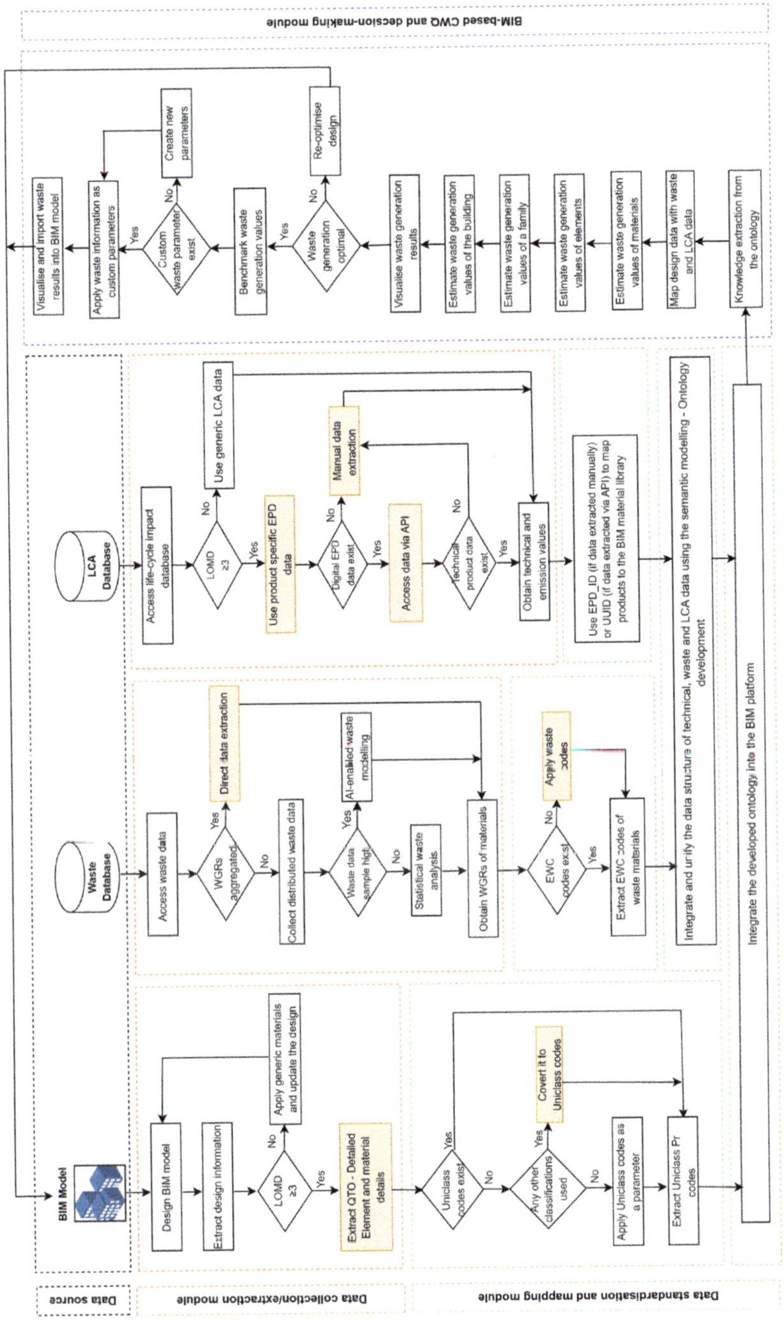

Figure 4.6 BIM-Based Semantic CWQ Framework

Source: Drafted by the author

information in the ontology will also lead to the development of reliant and complaint CW databases. The ontology becomes valuable only if the knowledge stored in it is extracted. The semantic SPARQL queries are used to retrieve explicitly or implicitly stored information in the ontology. The outputs retrieved from the queries are the primary results incorporated into the BIM model to measure the CW and its environmental impacts.

4.6.3 BIM-Based CWQ and Decision-Making Module

A BIM-based CWQ and decision-making module is introduced to enhance the dynamics between the BIM platform and the semantic CW data model. This component will provide a UI within the BIM environment for selecting an optimal design or re-selecting another design option should the initial design be deemed highly wasteful. The decision-making layer will allow users to CW waste generation levels for each material layer of an element individually, which will then be extended to other elements within an element family. Hence, the waste results of multiple element families that make up the building structure will provide the total waste generation value of the building (Figure 4.7). The GWP impacts and treatment methods of each waste material are also presented to allow users to select an optimised design that meets the project requirements.

The technical, waste, and carbon information of materials extracted from the ontology are presented as custom parameters within the BIM model. The quantification can be performed with one or multiple cycles until the user approves the specific material appropriate for the design. A new element family will be created as a final step, and the waste results will be presented in a tabular and graphical format. The quantified waste results incorporated into the BIM model allow the user to make waste-informed decisions from the initial stages of a project, thus overcoming the limitations of noninteroperable and static information flows between BIM and CW platforms. This decision layer further allows users to benchmark and compare multiple design options in the virtual environment before construction begins. So, the proposed semantic framework aims to enhance standardisation, consistency, and interoperability between BIM and CW platforms, further promising automation within the CWQ workflows.

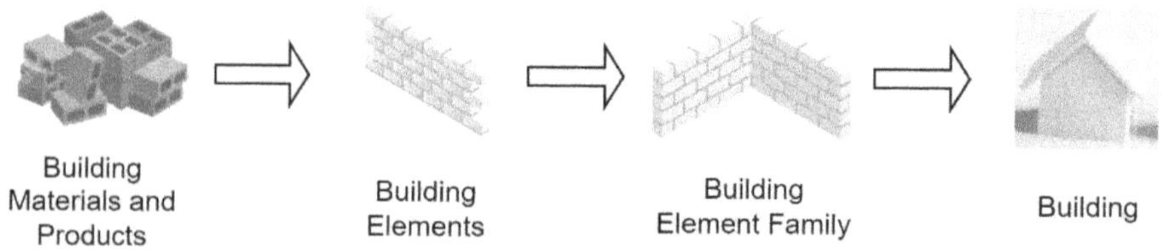

Figure 4.7 Building Levels Accounted for CWQ

Source: Drafted by the author

4.7 Summary

Aligning with the DSR approach, the framework was proposed. The framework developed is guided using the three steps, which include the (i) exploration of stakeholder perspectives through stakeholder interviews, (ii) identification of indicators supporting CW-informed decision-making via exploratory data analysis, and (iii) design of the framework. A critical realistic philosophical paradigm was used to design the research. The mixed method guided the three phases of the overall development process. The highlights of the chapter include the following.

Perspectives of stakeholders through semi-structured interviews:

- Integrating CW into decision-making in the design stages is vital to minimise waste generation.
- Advanced use of technologies and managing data heterogeneity is crucial to enhance the digitalisation and BIM compatibility of CW platforms.
- Proposing an impact-based CW assessment will allow the waste data to be integrated seamlessly into other environmental assessments like LCA, further easing communication.
- In addition to technological and process changes, communication and collaboration across the supply chain members, education and awareness, significant support from the client, and transparency are some of the vital elements that catalyse the CW reduction in the built environment.

CW indicators supporting decision-making:

- Based on the stakeholder's responses, 22 different types of information are ranked as indicators influencing CW-informed decision-making.
- The identified indicators are categorised as critical, high, medium, and low based on the frequency of occurrence. The influence of these indicators on CW decisions is decided based on the category in which they are placed. However, only the critical, high, and modest indicators are included in the study due to data and time constraints.

CWQ framework:

The three main elements are integrated into the framework to enable interoperability and automation across the CWQ workflow. It includes a unified data structure, standardised and consistent reporting, and BIM compatibility. To achieve the aim of the framework, three main modules are proposed, including (i) the data collection/extraction module, (ii) the data mapping and standardisation module, and (iii) the BIM-based CWQ and decision-making module. The proposed framework recognises the importance of data standardisation, mapping, and exchange between multiple platforms employed in the quantification process. The proposed framework seeks to enable integrated digital systems, ensure consistency in information flows, and produce sustainable results from ICT integration. Overall, the study places a high priority on the built environment to achieve interoperability and automation across the project workflows. The application of the proposed framework is demonstrated using a semantic CW data model (Ontology) and a BIM-integrated UI (Plugin) discussed in Chapters 5 and 6.

References

1. M.N. Saunders, P. Lewis, A. Thornhill and A. Bristow, Understanding research philosophy and approaches to theory development, in *Research methods for business students*, 8th edition, 2015. Harlow: Pearson Education, pp. 128–171. ISBN: 1-292-01662-0, 978–1–292–01662–7.
2. A. Bryman, *Social research methods*, 2016. Oxford University Press. ISBN: 0199689458.
3. B. Danermark, M. Ekström and J.C. Karlsson, *Explaining society: Critical realism in the social sciences*, 2019. Routledge. ISBN: 1351017810.
4. M. Saunders, P. Lewis and A. Thornhill, *Research methods for business students*, 2009. Pearson Education. ISBN: 0273716867.
5. A. Strauss and J. Corbin, *Basics of qualitative research*, 1990. Sage Publications. ISBN: 0803932502.
6. S.B. Merriam and E.J. Tisdell, *Qualitative research: A guide to design and implementation*, 2015. John Wiley & Sons. ISBN: 111900361X.
7. M. Sandelowski, Combining qualitative and quantitative sampling, data collection, and analysis techniques in mixed-method studies. *Research in Nursing & Health*, 2000. 23(3): pp. 246–255. https://doi.org/10.1002/1098–240X(200006)23:3%3C246::AID-NUR9%3E3.0.CO;2-H
8. A. Tashakkori and J.W. Creswell, The new era of mixed methods. *Journal of Mixed Methods Research*, 2007. 1(1): pp. 3–7. https://doi.org/10.1177/2345678906293042
9. U. Flick, *The Sage handbook of qualitative data collection*, 2017. Sage. ISBN: 1526416069.
10. C. Moustakas, *Phenomenological research methods*, 1994. Sage publications. ISBN: 145220747X.
11. W.C. Adams, Conducting semi-structured interviews, in *Handbook of practical program evaluation*, 2015: pp. 492–505. https://doi.org/10.1002/9781119171386.ch19
12. A. Kuzel, Sampling in qualitative inquiry, in *Doing qualitative research*, 1992: pp. 33–45.
13. S.E. Rabionet, How I learned to design and conduct semi-structured interviews: An ongoing and continuous journey. *Qualitative Report*, 2011. 16(2): pp. 563–566. https://doi.org/10.46743/2160–3715/2011.1070
14. G. Sharma, Pros and cons of different sampling techniques. *International Journal of Applied Research*, 2017. 3(7): pp. 749–752.
15. I. Etikan, S.A. Musa and R.S. Alkassim, Comparison of convenience sampling and purposive sampling. *American Journal of Theoretical and Applied Statistics*, 2016. 5(1): pp. 1–4. https://10.11648/j.ajtas.20160501.11
16. V. Braun and V. Clarke, *Thematic analysis*, 2012. American Psychological Association. ISBN: 1433810050.
17. C. Grbich, *Qualitative data analysis: An introduction*, 2012. Sage. ISBN: 1446202968.
18. M.E. Kiger and L. Varpio, Thematic analysis of qualitative data: AMEE guide no. 131. *Medical Teacher*, 2020. 42(8): pp. 846–854. https://doi.org/10.1080/0142159X.2020.1755030
19. V. Braun and V. Clarke, Using thematic analysis in psychology. *Qualitative Research in Psychology*, 2006. 3(2): pp. 77–101.

20. C. Cassell and G. Symon, *Essential guide to qualitative methods in organizational research*, 2004. SAGE Publications Ltd. ISBN: 144622418X.
21. J. Ritchie and L. Spencer, Qualitative data analysis for applied policy research, in *Analyzing qualitative data*. 2002, Routledge. pp. 173–194. ISBN: 9780203413081.
22. S. Nagapan, I.A. Rahman, A. Asmi, A.H. Memon and I. Latif, Issues on construction waste: The need for sustainable waste management, in *2012 IEEE colloquium on humanities, science and engineering (CHUSER)*, 2012. IEEE. https://doi.org/10.1109/CHUSER.2012.6504333
23. M. Osmani, Construction waste minimization in the UK: Current pressures for change and approaches. *Procedia-Social and Behavioral Sciences*, 2012. 40: pp. 37–40. https://doi.org/10.1016/j.sbspro.2012.03.158
24. H. Yuan, Critical management measures contributing to construction waste management: Evidence from construction projects in China. *Project Management Journal*, 2013. 44(4): pp. 101–112. https://doi.org/10.1002/pmj.21349
25. A. Bakchan, K.M. Faust and F. Leite, Seven-dimensional automated construction waste quantification and management framework: Integration with project and site planning. *Resources, Conservation and Recycling*, 2019. 146: pp. 462–474. https://doi.org/10.1016/j.resconrec.2019.02.020
26. B.C. Guerra, F. Leite and K.M. Faust, 4D-BIM to enhance construction waste reuse and recycle planning: Case studies on concrete and drywall waste streams. *Waste Management*, 2020. 116: pp. 79–90. https://doi.org/10.1016/j.wasman.2020.07.035
27. J. Won, J.C. Cheng and G. Lee, Quantification of construction waste prevented by BIM-based design validation: Case studies in South Korea. *Waste Management*, 2016. 49: pp. 170–180. https://doi.org/10.1016/j.wasman.2015.12.026
28. J.C. Cheng and L.Y. Ma, A BIM-based system for demolition and renovation waste estimation and planning. *Waste Management*, 2013. 33(6): pp. 1539–1551. https://doi.org/10.1016/j.wasman.2013.01.001
29. Z. Wu, T. Ann, L. Shen and G. Liu, Quantifying construction and demolition waste: An analytical review. *Waste Management*, 2014. 34(9): pp. 1683–1692. https://doi.org/10.1016/j.wasman.2015.12.026
30. O.O. Akinade and L.O. Oyedele, Integrating construction supply chains within a circular economy: An ANFIS-based waste analytics system (A-WAS). *Journal of Cleaner Production*, 2019. 229: pp. 863–873. https://doi.org/10.1016/j.jclepro.2019.04.232
31. A. Bakshan, I. Srour, G. Chehab and M. El-Fadel, A field based methodology for estimating waste generation rates at various stages of construction projects. *Resources, Conservation and Recycling*, 2015. 100: pp. 70–80. https://doi.org/10.1016/j.resconrec.2015.04.002
32. Deloitte, Resource efficient use of mixed wastes, improving management of construction and demolition waste – Final report, in *Prepared for the European Commission*, 2017. DG ENV. https://environment.ec.europa.eu/system/files/2021–01/resource_efficient_uses_mixed_waste_Final_Report.pdf. Last access: 07 December 2024.
33. M. Bilal, L.O. Oyedele, J. Qadir, K. Munir, S.O. Ajayi, O.O. Akinade, H.A. Owolabi, H.A. Alaka and M. Pasha, Big Data in the construction industry: A review of present status, opportunities, and future trends. *Advanced Engineering Informatics*, 2016. 30(3): pp. 500–521. https://doi.org/10.1016/j.aei.2016.07.001
34. Z. Yu, H. Peng, X. Zeng, M. Sofi, H. Xing and Z. Zhou, Smarter construction site management using the latest information technology, in *Proceedings of the institution of civil engineers-civil engineering*, 2018. Thomas Telford Ltd. https://doi.org/10.1680/jcien.18.00030
35. The Institution of Engineering and Technology, *Digitisation for construction product manufacturers – A plain language guide*, 2021. https://www.theiet.org/impact-society/policy-and-public-affairs/digital-futures-policy/reports-and-papers/digitisation

-for-construction-product-manufacturers-a-plain-language-guide. Last access: 07 December 2024.

36. R. Kitchin, *The data revolution: Big data, open data, data infrastructures and their consequences*, 2014. Sage. ISBN: 9781473909472.
37. S. Sivashanmugam, S. Rodriguez, F.P. Rahimian, F. Elghaish and N. Dawood, Enhancing information standards for automated construction waste quantification and classification. *Automation in Construction*, 2023. 152: p. 104898. https://doi.org/10.1016/j.autcon.2023.104898
38. S. Azhar, Building information modeling (BIM): Trends, benefits, risks, and challenges for the AEC industry. *Leadership and Management in Engineering*, 2011. 11(3): pp. 241–252. https://doi.org/10.1061/(ASCE)LM.1943–5630.0000127
39. C.M. Eastman, P. Teicholz, R. Sacks and K. Liston, *BIM handbook: A guide to building information modeling for owners, managers, designers, engineers and contractors*, 2011. John Wiley & Sons. ISBN: 978-1-119-28753-7.
40. NBS, *Digital construction report 2021 – Incorporating the BIM report*, 2021. https://www.thenbs.com/digital-construction-report-2021/. Last access: 07 December 2024.
41. A. Nagalli, Estimation of construction waste generation using machine learning, in *Proceedings of the institution of civil engineers-waste and resource management*, 2021. Thomas Telford Ltd. https://doi.org/10.1680/jwarm.20.00019
42. N. Rodrigo, H. Omrany, R. Chang and J. Zuo, Leveraging digital technologies for circular economy in construction industry: A way forward. *Smart and Sustainable Built Environment*, 2024. 13(1): pp. 85–116. https://doi.org/10.1108/SASBE-05-2023-0111
43. C.T. Formoso, L. Soibelman, C. De Cesare and E.L. Isatto, Material waste in building industry: Main causes and prevention. *Journal of Construction Engineering and Management*, 2002. 128(4): pp. 316–325. https://doi.org/10.1061/(ASCE)0733–9364(2002)128:4(316)
44. G. Hobbs, K. Adams and M. Blackwell, Understanding and predicting construction waste, in *Proceedings of the institution of civil engineers-waste and resource management*, 2011. ICE Publishing. https://doi.org/10.1680/warm.2011.164.4.239
45. W. Lu, H. Yuan, J. Li, J.J. Hao, X. Mi and Z. Ding, An empirical investigation of construction and demolition waste generation rates in Shenzhen city, South China. *Waste Management*, 2011. 31(4): pp. 680–687. https://doi.org/10.1016/j.wasman.2010.12.004
46. CLC and The Green Construction Board, *The routemap for zero avoidable waste in construction*, 2021. https://www.constructionleadershipcouncil.co.uk/wp-content/uploads/2021/07/ZAW-Interactive-Routemap-FINAL.pdf. Last access: 07 December 2024.
47. M. Kucukvar, G. Egilmez and O. Tatari, Evaluating environmental impacts of alternative construction waste management approaches using supply-chain-linked life-cycle analysis. *Waste Management & Research*, 2014. 32(6): pp. 500–508. https://doi.org/10.1177/0734242X14536457
48. H. Wu, H. Duan, J. Wang, T. Wang and X. Wang, Quantification of carbon emission of construction waste by using streamlined LCA: a case study of Shenzhen, China. *Journal of Material Cycles and Waste Management*, 2015. 17: pp. 637–645. https://doi.org/10.1007/s10163-015-0404-9
49. J. Monahan and J.C. Powell. An embodied carbon and energy analysis of modern methods of construction in housing: A case study using a lifecycle assessment framework. *Energy and Buildings*, 2011. 43(1): pp. 179–188. https://doi.org/10.1016/j.enbuild.2010.09.005
50. Defra, *Our waste, our resources: A strategy for England*, 2018. https://assets.publishing.service.gov.uk/government/uploads/system/uploads/attachment_data/file/765914/resources-waste-strategy-dec–2018.pdf. Last access: 02 April 2023.
51. A. Naneva, M. Bonanomi, A. Hollberg, G. Habert and D. Hall, Integrated BIM-based LCA for the entire building process using an existing structure for cost estimation in the Swiss context. *Sustainability*, 2020. 12(9): p. 3748. https://doi.org/10.3390/su12093748

52. E. Palumbo, B. Soust-Verdaguer, C. Llatas and M. Traverso, How to obtain accurate environmental impacts at early design stages in BIM when using environmental product declaration. A method to support decision-making. *Sustainability*, 2020. 12(17): p. 6927. https://doi.org/10.3390/su12176927
53. B. Swartout, R. Patil, K. Knight and T. Russ, Toward distributed use of large-scale ontologies. *Proc. of the tenth workshop on knowledge acquisition for knowledge-based systems*, 1996. https://cdn.aaai.org/Symposia/Spring/1997/SS-97-06/SS97-06-018.pdf
54. Commission of the European Communities, *Taking sustainable use of resources forward: A thematic strategy on the prevention and recycling of waste*, 2005. https://eur-lex.europa.eu/LexUriServ/LexUriServ.do?uri=COM:2005:0666:FIN:EN:PDF. Last access: 07 December 2024.
55. K. Safari and H. AzariJafari, Challenges and opportunities for integrating BIM and LCA: Methodological choices and framework development. *Sustainable Cities and Society*, 2021. 67: pp. 102728. https://doi.org/10.1016/j.scs.2021.102728
56. S. Deep, T. Gajendran and M. Jefferies, A systematic review of 'enablers of collaboration' among the participants in construction projects. *International Journal of Construction Management*, 2021. 21(9): pp. 919–931. https://doi.org/10.1080/15623599.2019.1596624
57. M. Murray and D. Langford, *Construction reports 1944–98*, 2008. ISBN: 978-1-405-14755-2.
58. M. Oraee, M.R. Hosseini, D.J. Edwards, H. Li, E. Papadonikolaki and D. Cao, Collaboration barriers in BIM-based construction networks: A conceptual model. *International Journal of Project Management*, 2019. 37(6): pp. 839–854. https://doi.org/10.1016/j.ijproman.2019.05.004
59. M.F. Dulaimi, F.Y. Ling and A. Bajracharya, Organizational motivation and inter-organizational interaction in construction innovation in Singapore. *Construction Management and Economics*, 2003. 21(3): pp. 307–318. https://doi.org/10.1080/0144619032000056144
60. X. Xue, R. Zhang, L. Wang, H. Fan, R.J. Yang and J. Dai, Collaborative innovation in construction project: A social network perspective. *KSCE Journal of Civil Engineering*, 2018. 22: pp. 417–427. https://doi.org/10.1007/s12205-017-1342-y
61. HM Government, *Construction 2025*, 2013. https://assets.publishing.service.gov.uk/media/5a7b7ca140f0b62826a03f2c/bis-13–955-construction-2025-industrial-strategy.pdf. Last access: 07 December 2024.
62. P. Ghisellini, C. Cialani and S. Ulgiati, A review on circular economy: The expected transition to a balanced interplay of environmental and economic systems. *Journal of Cleaner Production*, 2016. 114: pp. 11–32. https://doi.org/10.1016/j.jclepro.2015.09.007
63. K.T. Adams, M. Osmani, T. Thorpe and J. Thornback, Circular economy in construction: Current awareness, challenges and enablers, in *Proceedings of the institution of civil engineers-waste and resource management*, 2017. Thomas Telford Ltd. https://doi.org/10.1680/jwarm.16.00011
64. N. Udawatta, J. Zuo, K. Chiveralls and G. Zillante, Attitudinal and behavioural approaches to improving waste management on construction projects in Australia: Benefits and limitations. *International Journal of Construction Management*, 2015. 15(2): pp. 137–147. https://doi.org/10.1080/15623599.2015.1033815
65. S.C. e Silva, F. Bradley and C.M. Sousa, Empirical test of the trust-performance link in an international alliances context. *International Business Review*, 2012. 21(2): pp. 293–306. https://doi.org/10.1016/j.ibusrev.2011.03.006
66. X. Zhao, Stakeholder-associated factors influencing construction and demolition waste management: A systematic review. *Buildings*, 2021. 11(4): p. 149. https://doi.org/10.3390/buildings11040149
67. E. Küçüksayraç, D. Keskin and H. Brezet, Intermediaries and innovation support in the design for sustainability field: Cases from the Netherlands, Turkey and the United Kingdom. *Journal of Cleaner Production*, 2015. 101: pp. 38–48. https://doi.org/10.1016/j.jclepro.2015.03.078

68. M. Sandelowski, C.I. Voils and G. Knafl, On quantitizing. *Journal of Mixed Methods Research*, 2009. 3(3): pp. 208–222. https://doi.org/10.1177/1558689809334210
69. M. Sandelowski, Real qualitative researchers do not count: The use of numbers in qualitative research. *Research in Nursing & Health*, 2001. 24(3): pp. 230–240. https://doi.org/10.1002/nur.1025
70. O. Disney, M. Roupé, M. Johansson and A.D. Leto, Embracing BIM in its totality: A total BIM case study. *Smart and Sustainable Built Environment*, 2022. 13(3): pp. 512–531. https://doi.org/10.1108/SASBE-06-2022-0124
71. A. Kamari, B.M. Kotula and C.P.L. Schultz, A BIM-based LCA tool for sustainable building design during the early design stage. *Smart and Sustainable Built Environment*, 2022. 11(2): pp. 217–244. https://doi.org/10.1108/SASBE-09-2021-0157
72. S. Sivashanmugam, S. Rodriguez, F.P. Rahimian and N. Dawood, Maximising the construction waste reduction potential-barriers and catalysts, in *Proceedings of the institution of civil engineers-civil engineering*, 2022. Thomas Telford Ltd. https://doi.org/10.1680/jcien.22.00163
73. R. Santos, A.A. Costa, J.D. Silvestre and L. Pyl, Integration of LCA and LCC analysis within a BIM-based environment. *Automation in Construction*, 2019. 103: pp. 127–149. https://doi.org/10.1016/j.autcon.2019.02.011
74. S. Sivashanmugam, S. Rodriguez Trejo and F.P. Rahimian, BIM-integrated semantic framework for construction waste quantification and optimisation. *Automation in Construction*, 2024. 168: p. 105842. https://doi.org/10.1016/j.autcon.2024.105842
75. Construction Leadership Council, *Zero avoidable waste in construction*, 2020. https://www.constructionleadershipcouncil.co.uk/wp-content/uploads/2016/05/ZAW-Report-Final-Draft-25-February–2020.pdf. Last access: 15 May 2023.
76. NBS, *Uniclass – Unified construction classification*, 2022. https://uniclass.thenbs.com/. Last access: 12 March 2023.
77. Environment Agency, *Guidance on the classification and assessment of waste* (1st Edition v1.2.GB) – Technical Guidance WM3, 2021. https://assets.publishing.service.gov.uk/government/uploads/system/uploads/attachment_data/file/1021051/Waste_classification_technical_guidance_WM3.pdf. Last access: 02 June 2023

5 Product Circularity Ontology – Design, Development, and Validation

5.1 Introduction

The application of the proposed CWQ framework is demonstrated using a semantic CW data model. The proposed semantic data model is a metadata schema that provides explicit specifications of CW entities and relationships among them. The standard semantics established within the model ease understanding of the CW information and knowledge sharing within the built environment. The semantic CW data model is prefixed as PROduct CIRcularity Ontology (PRODCIRO). The SKEM approach supported the design and development of PRODCIRO. The Specification, conceptualisation, formalisation, and implementation of the ontology are discussed in SKEM in seven steps. The steps include domain and scope determination, reuse of existing ontologies, defining the class and class hierarchies, defining the properties of the class – slots, defining the facets of the slots, and creating instances. These seven steps are adopted to model PRODCIRO. The syntactic and semantic consistency of the ontology is evaluated using the Pellet reasoner. The knowledge from the ontology is extracted using the SPARQL queries. The output of the queries demonstrates the efficiency of PRODCIRO design in answering the required questions and hypotheses and inferring logical and meaningful insights from the relationships established between several entities. The outputs of the literature review, thematic data analysis, and exploratory data analysis facilitated the PRODCIRO development.

5.2 PRODCIRO Design and Development

The literature and interview results emphasise the need for an ontology with a standardised, unified, and consistent CW data structure. Hence, the functionalities of BIM and SWT are utilised to develop an ontology that aids in quantifying CW and informs optimal material choices from early design stages. As stated by [1], the applications of SWT in the AECO industry benefit existing technologies like BIM to enhance interoperability and create collaborative information management across various domains. Similarly, the ontology proposed in the research aims to enhance seamless information flows between BIM and CW platforms. There is no specific method to model an ontology, but modelling is influenced by designers' understanding and domain applications of the ontology [2]. Although existing

DOI: 10.1201/9781003610762-5

OEMs are almost similar, the research adopts the SKEM approach for ontology development (discussed in Section 3.4.7). The seven steps of SKEM adopted to model PRODCIRO include the following:

i) Determine the domain and scope of the ontology
ii) Consider the reuse of ontologies
iii) Enumerate important terms in the ontology
iv) Define the classes and class hierarchies
v) Define the properties of classes – slots
vi) Define the facets of the slots
vii) Create instances

The seven steps are grouped as pre-development, development, and post-development phases. The pre-development phase involves preliminary ground works conducted to determine the domain of the ontology, the scope the ontology is expected to cover, and the questions or hypotheses the ontology is anticipated to answer. This stage allows the designer to define the overall purpose of the ontology. The outputs from the pre-development phase act as input for ontology design and development. The process associated with the pre-development phase is called specification. The second group, the development phase, encompasses Steps 2–7 of SKEM. The actual design and development of the ontology happens at this stage. The conceptualisation of important entities that will be used in the ontology to answer the framed questions is proposed at this stage.

Furthermore, the conceptualised entities are formalised and implemented into the actual development process. The primary entities within the ontology will be classes (also called concepts, as explained in Section 3.4.4), object properties, data properties, and instances (also called individuals, as explained in Section 3.4.4). The relationship established between these entities answers the ontology questions. Furthermore, the post-development stage involves the maintenance of the ontology across various scenarios. Since the scope of the research is to prove the importance of using the ontological model to facilitate CW data integration from heterogeneous data sources, the tasks associated with the pre-development and development phases are only included in the research.

5.2.1 *Determine the Domain and Scope of the Ontology*

The primary step in the ontology development process is to define the domain and scope of the ontology, considering the existing scenarios and future applications, called the pre-development phase. A few primary questions must be answered before ontology development [2], which also acts as a technique for establishing ontology specification requirements. The questions the ontology should answer are developed based on the outcomes of the literature review and semi-structured interviews.

Q1) *What is the purpose of the ontology?* – The ontology aims to develop an interoperable BIM-integrated ontological model with a unified and consistent data structure to enhance CWQ from the early design stages.

Q2) *Who will use and maintain the ontology/who are the end-users?* – The intended users of the ontology are the design team members, including designers, architects, and sustainability professionals who aim to develop preventive CWM measures in the design phase.

Q3) *For what types of questions should the information in the ontology provide answers/what are its intended uses?* – The information within the ontology should help stakeholders make informed waste decisions and optimise materials and designs, thereby facilitating the built environment in minimising CW generation, maximising building circularity, and reducing WLC carbon levels.

Q4) *What are the domains that the ontology will cover?* – The ontology covers the CW domain of building projects in the UK. It includes technical, environmental, and generic elements of CW that are crucial for making informed decisions in the design phases (obtained from Section 4.5.2).

Table 5.1 Knowledge Sources for Ontology Modelling

Data source	*Data type*	*Source type*	*Data extracted*	*Publisher*	*Reference*
BRE Waste Benchmark Data (2021)	Database	Closed data	WGR of building materials during the construction phase	BRESmartWaste	NA[1]
EPD certificates	PDF/XML documents	Open (or licensed) data	Embodied carbon emissions and EoL scenarios of building materials/products	GreenBook Live/Eco Portal	NA[2]
Uniclass	Website	Open data	Codes used to store, structure, communicate, and transfer project information across various AECO domain	NBS	[3]
Zero Avoidable Waste in Construction	Guidance document	Open data	Guidelines for employing EPD-based resource use and EoL results in the CW assessment	CLC and The Green Construction Board	[4]
Whole-life carbon assessment for the built environment	Guidance document	Open data	Guidelines for conducting whole life-cycle analysis and making generic assumptions in case of lack of specific data.	Royal Institute of Chartered Surveyors (RICS)	[5]
EWC Technical Guidance	Guidance document	Open data	Six-digit codes for CW reporting that align with legal standards	Environment Agency	[6]

NA[1] – Since this is a closed dataset, no reference can be provided.

NA[2] – Multiple EPDs exist, so referencing would not be appropriate. The IDs of selected EPDs are provided in Chapter 6.

Source: Drafted by the author

Q5) *What are the sources of knowledge used to develop the ontology?* – The BIM material database, BRESmartWaste data, EPDs, and Uniclass codes are primary sources of knowledge for ontology modelling (Table 5.1).

In addition to the fundamental questions, the CQs are answered to underline the lower-level specifications of the ontology. They are natural language questions that the ontology to be built should be able to answer. The CQs are content-specific and vary with the domain and functional requirements of the ontology [7]. In this research, the CQs to be answered include, but are not limited to, the following:

1) What type of projects will be a case for ontology modelling? – Building projects.
2) What technical factors of materials affect waste generation? – Service life, product size, product density, and product mass.
3) What type of waste information will facilitate CW-informed decisions? – Secondary materials usage, reusability, recyclability, energy recovery, landfilling levels of waste materials, and EWC.
4) What carbon factors are needed to estimate the carbon levels associated with waste? – Material-related factors.
5) Which mapping system is adopted to allow the entities in the knowledge base to align with BIM libraries? – Uniclass Product (Pr) and Element and Functions (EF) codes.
6) How should optimised products be retrieved from the list? – Based on its application, technical attributes, CW quantities, embodied carbon levels, and EoL scenarios.

Judging from the list of questions, the ontology will include information on various material types, their technical characteristics, environmental impacts, and WGRs to enable accurate quantification and optimisation. Though the scope definition and ontology development are iterative, these initial questions will help the designer reduce redundancies and identify the essential information required for modelling.

5.2.2 *Consider Reuse of Ontologies*

After the domain specification, the fundamental step of conceptualisation is to check for any existing general or domain ontologies that can be reused for modelling [2]. The ontology reuse is a process in which the existing ontologies are used to model the new ontologies. The ontologies can be reused as a whole or individual terms/statements depending on their application in the selected domain. In such contexts, the knowledge generated from existing ontologies serves as an input to generate new ontologies. The ontology reuse facilitates the development of knowledge-based systems, in which their domain knowledge and representation are independent of the software systems and allow seamless sharing and reusing across various software systems. The interoperability achieved using ontology reuse further improves the quality and application of the models. Besides, utilising the existing ontology components on the web would reduce the cost of developing new ontologies and information redundancies [2, 8].

Several ontologies have been developed in the AECO domain to support sustainability assessment, building performance analysis, energy modelling, and BIM-GIS integration. The Linked Building Data community has developed several standard ontologies like Building Topology Ontology and ifcOWL to enhance standardisation across ontologies in the AECO domain [9, 10]. Despite the existence of ontologies, the lack of generalisation and application dependence are significant shortcomings that reduce their sharing and reusing capabilities. Also, the standard ontologies like ifcOWL are more extensive, thus increasing the complexity of utilising them for new ontology models [11].

Other two ontologies, Building Materials Ontology (BMO) [12] and Materials Properties Ontology (MAT) [13], are developed to provide the explicit relationship between elements, materials, and their properties aligning with BIM, which is the prospective relation anticipated to obtain from the current research. Although multiple material properties are included in the BMO, they all lean towards technical aspects while excluding critical environmental factors. Thus, the ontology cannot be directly reused in the research due to a lack of CW and carbon entities. However, the concepts and attributes stated in the BMO, MAT, Quantities, Units, Dimensions, and Types (QUDT) ontologies are incorporated into the modelling process. Some include 'BMO:MaterialType', 'BMO:MaterialProperty', 'BMO:hasThickness', 'QUDT:Unit', and the class and sub-classes of 'MAT:Insulation material'.

Apart from direct extraction, other concepts used in the existing ontologies are modified according to the research needs. For example, the 'MAT: Brick masonry' and 'MAT: Concrete masonry' come under the super-class 'MAT:Masonry', but in this research, the brick and concrete act as an individual class, and its usage, such as masonry units, slabs, floors, and decks, acts as sub-class of the material. So, these terms are modified to fit the ontology. Therefore, these concepts act as critical inputs for ontological model development, which can also be merged to integrate with BIM and IFC processes. To mention a few, the material properties 'density' and 'material volume' provide the information required to calculate the mass of the material, which in turn provides CW quantities. So, these terms are included in the PRODCIRO.

5.2.3 *Enumerate Important Terms in the Ontology*

The output of 'knowledge identification' provides a glossary of important terms that fit the domain and scope of the ontology [14]. The terms include taxonomies of concepts, attributes, and individuals required to fulfil the ontology specification requirements. In this research, the factors affecting CW generation, including building type, material type, materials' dimensional features, density, service life, WGR, embodied carbon, the quantity of raw and secondary materials, EoL scenarios, applications via Uniclass codes, EWCs, and manufacturer details are some of the essential terms gathered to facilitate ontology modelling. The above-listed factors depend on the information collected from the literature review, domain-related knowledge sources, and exploratory data analysis (Chapters 2 and 4). These terms are collectively called non-ontological resources (NORs) in the NeOn ontology engineering method [15]. The NORs is a knowledge source whose semantics are not yet formalised by the ontology. These knowledge sources are heterogeneous, restraining their direct application across the software systems. So, the collected knowledge is organised and structured to develop a conceptual knowledge model that facilitates ontology modelling. This process of describing knowledge about a domain is labelled as conceptualisation [14]. The conceptualisation provides informal knowledge extracted and generalised from experience, observation, and introspection [16]. Similarly, the vocabulary of BIM, material, waste, Uniclass, and EPDs provide an informal CW knowledge base.

As [2] suggested, informal knowledge provides an outline for the developer to model ontologies. Nevertheless, it also provides flexibility to update or change these terms as the development continues, given that the new terms fit the domain scope. Therefore, the listed factors in this research could be modified in the final ontology depending on the data consistencies within the CW domain.

5.2.4 Define the Classes and the Class Hierarchy

The conceptual model with an identified hierarchy of classes and its associated properties are structured into the ontology. This process is called formalisation. Formalisation, a process of transforming informally defined knowledge in a conceptual model to a formalised ontology language, is aimed at enhancing the computation factors of the ontologies [14]. Informal knowledge is represented using logical language, which is made up of classes and relations. This logical language defines how the specific domain represents the information and its relations. Formal ontologies overcome the data heterogeneity issues by enabling a consistent and standard representation of information. The tasks associated with the formalisation are discussed in the following sections.

Naming and Prefixes:

It is important to note that, in the semantic web, each item or entity is a resource identified by a URI/IRI identifier. The URI allows the ontology to address the resources across the web uniquely. However, the URI is compressed to a specific character set, which lacks its universal applications. With increasing growth in web data, a rich and universal character set is needed, where the IRI plays a role. The IRI extends the syntax of URI, using the characters from the universal character set, which enables internationalised and uniformised web addressing. Hence, IRIs complement URIs by permitting the use of non-URI characters. So, the names of classes, properties, and instances used in the research are IRIs. The IRIs could also be written in a short form using a prefix and a local name in the *prefix:local name* format. The prefix is a short form that expands to an IRI, and the local name is the name given to the entity. To illustrate, in the IRI, 'http://tees.ac.uk/EPD_Data/Lifecycle_Impact_Assessment_Results#BuildingType' http://tees.ac.uk/EPD_

Data/Lifecycle_Impact_Assessment_Results# is the IRI, and *BuildingType* is the entity. This is written as PRODCIRO:BuildingType, where PRODCIRO is the prefix and *BuildingType* is the local name of the entity. Figure 5.1 presents the prefixes and corresponding IRIs used in the ontology. The BMO, MAT, and QUDT represent the IRIs taken from the respective ontologies (discussed in Section 5.2.2). The PRODCIRO is the primary prefix name given to the IRIs created within the ontology. The *local name* will be used in the text and tables, and the *prefix:local name* will be used in the figures.

It is also emphasised in the web that there are no embedded spaces in the identifiers. To name a few, the terms 'Building Material' and ' has manufacturer' will not be used on the web. Instead, the resources with multiple names are identified without spaces in between them and capitalising each word. So, 'Building Material' becomes 'BuildingMaterial' and 'has manufacturer' becomes 'hasManufacturer'. The conversion of the entity names is called InterCaps or CamelCase conversion [17]. Therefore, each item discussed in the following sections is a 'CamelCase prefixed local name' of the IRI.

5.2.4.1 *Class-Hierarchy Development*

According to the SKEM method, Sections 5.2.3–5.2.7 are intertwined and iterative, thus leading to continuous updates of the classes and attributes and

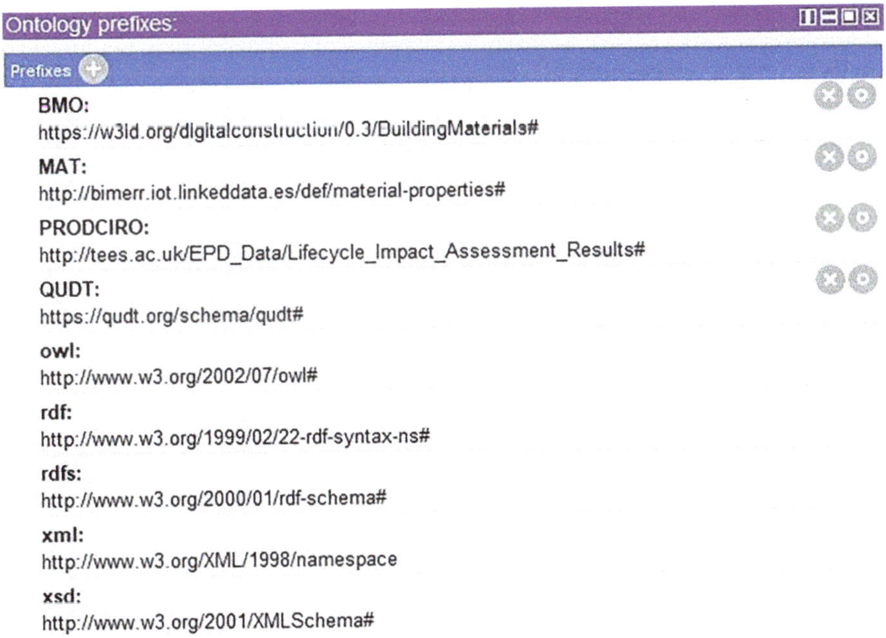

Figure 5.1 Ontology Prefixes and IRIs

Source: Drafted by the author

their adoption into a formal ontology. There are several viable approaches for class-hierarchy development. This research borrows the three approaches suggested by [18] to develop a class hierarchy. It includes a top-down approach, a bottom-up approach, and a middle-out/combination approach. The top-down approach starts with more generalised ontology domain classes and continues with the subsequent specialised classes. For example, *BuildingElements* is a generalised class; element types like *Walls, Roofs, Floors, Beams, and Columns* are specialised classes. So, in the top-down approach, the generalised class is defined first, which will then be followed by its specialised classes. The generalised classes will act as a super-class, and specialised classes will be defined as sub-classes. This process is vice versa in the bottom-up approach. The specialised classes (element types) are defined first and grouped later in a more general class (*BuildingElements*).

Unlike the top-down and bottom-up approaches, the middle-out approach puts no restrictions on defining a class. It gives the flexibility to define more significant classes first and then generalise and specialise them appropriately according to the ontology specification. So, the developer starts with a few super-classes like *BuildingElements*, and the low-level classes like *ExternalWalls*, *InternalWalls,* and *PartitionWalls* are later connected through a middle-level class called *Walls*. In this case, the low-level class is an immediate sub-class of the middle-level class, and the middle-level class is an immediate sub-class of the super-class. Although no single method inherently performs better than others, the choice depends on the ontology developers [2]. Therefore, the middle-out combined approach is selected for the research. It is because of its flexibility in implementation while taking the edges of top-down and bottom-up approaches. Besides, the developers widely prefer this method for its benefits across accuracy, granularity, and reduced reworks [2]. Despite the type of approach used, the first step is to outline the classes.

5.2.4.2 PRODCIRO Classes and Their Hierarchies

The entities defined within the ontology are characterised as asserted or inferred. The asserted views present the assumptions asserted by the user in OWL files. Meanwhile, the inferred view shows the exact position of the classes semantically. The reasoner is used to infer the implicit facts within the ontology. The asserted view is a default view in Protégé, while the inferred views will be void until a reasoner is used. The inferred class hierarchy is also a method of presenting the reasoner outcomes. The Protégé environment, the selected tool for ontology development in the research (discussed in Section 3.4.6), allows the user to display class-hierarchy views as asserted or inferred. So, the views displayed in this section are inferred class-hierarchy views.

The general classes of PRODCIRO include *EPDProductType, EPDProductData, BuildingWasteMaterialType, BuildingType, BuildingElements,*

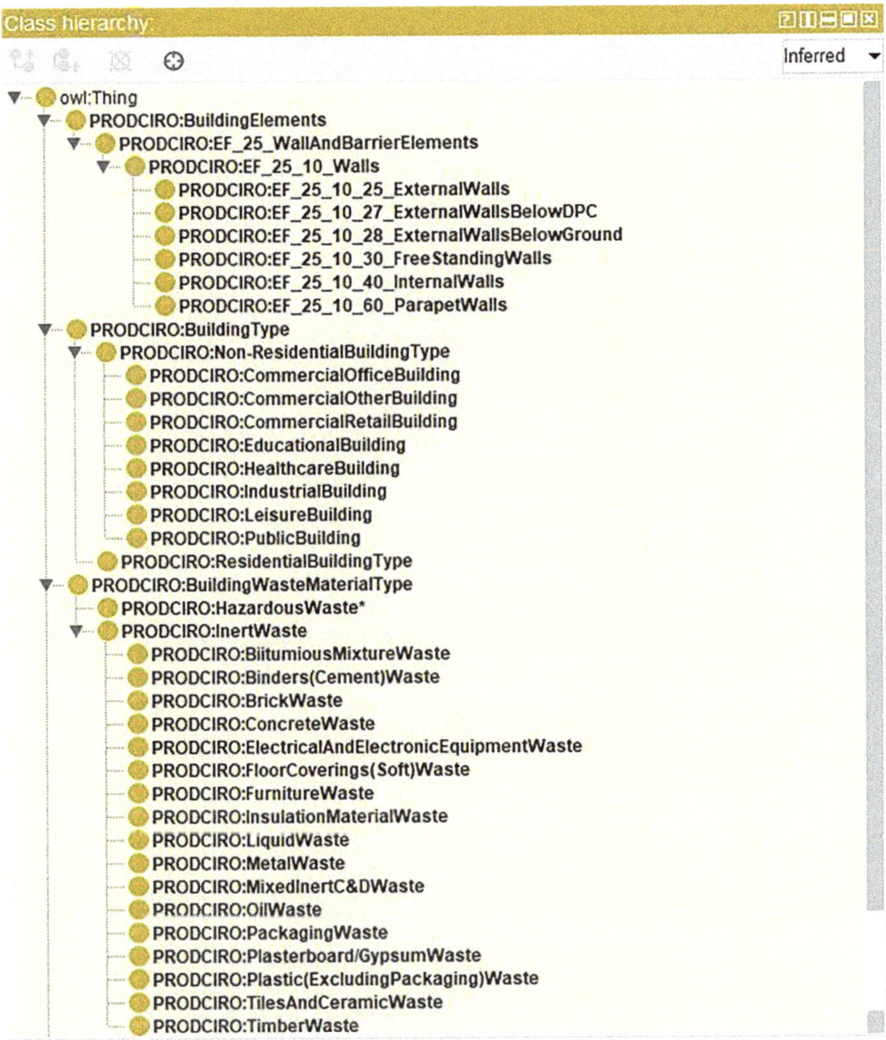

Figure 5.2 Class Hierarchy of PRODCIRO – Part A

Source: Drafted by the author

EuropeanWasteCatalogue_EWC, UniclassPrCodes, and *Unit*. These are the top classes of PRODCIRO (Figures 5.2 and 5.3). The top classes have middle-level and low-level classes. All these class types are rooted under *owl:Thing*. To illustrate, *BuildingType* has two middle-level classes (*ResidentialBuilding-Type* and *Non-ResidentialBuildingType*). The latter type is further classified as

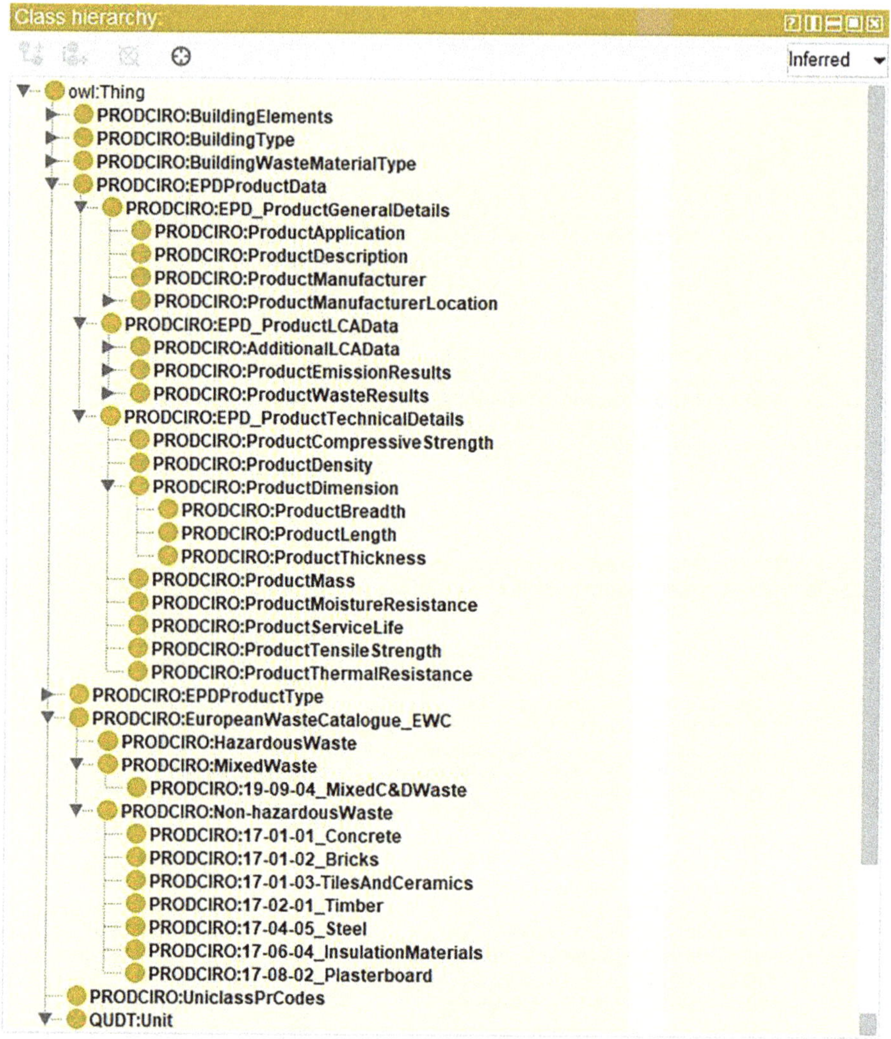

Figure 5.3 Class Hierarchy of PRODCIRO – Part B

Source: Drafted by the author

CommercialBuildings, *IndustrialBuildings*, and so on. This is the final categorisation under the *BuildingType* classes, which are called low-level classes. An overall view of the type of classes, the type of information they hold and the justification for their inclusion are presented in Table 5.2. A detailed description of each class, its categories, and its relationship with other classes is discussed in the following sections.

Table 5.2 List of PRODCIRO Classes, Type of Information in the Classes, and Justification for Its Inclusion

Classes	*Type of information in the classes*	*Justification for its inclusion in the ontology*
EPDProductType	The EPDs of each building product come under this class.	It is one of the main classes added to the ontology to classify building products. It links with the generic, technical, and environmental data added to the *EPDProductData* class.
EPDProductData	A primary class that holds generic, technical, carbon, secondary material usage, and expected waste treatment scenario details of building products extracted from the EPDs.	The data extracted from this class is integrated with waste data from the *BuildingWasteMaterialType* class to quantify, classify, and optimise materials based on CW impact results.
BuildingWaste Material Type	Information on the material composition of CW recorded in the waste reports is added to this class.	One of the main classes used to classify CW. The classification is based on their material composition. The data extracted from this class is integrated with waste data from the *EPDProductData* class to quantify, classify, and optimise materials based on CW impact results.
BuildingType	Type of building usages: residential and non-residential (commercial, industrial, educational, healthcare, leisure, and public) buildings.	According to BRESmartWaste reports, building typology is one of the main factors influencing waste generation during the construction/installation phase of the building. So, this class is incorporated into the CWQ.
BuildingElements	This is a primary class with a collection of building elements. The building elements are classified according to the Uniclass EF classification.	Each product has a unique application across various building elements. So, this class is used to select and match the products (within the *EPDProductType* class) applicable to a specific element (within the *BuildingElements* class).
EuropeanWaste Catalogue_EWC	Contain EWC codes of each waste material reported in *BuildingWasteMaterialType*.	These codes are used to select an effective treatment method and allow safe disposal of waste materials, aligning with the legal waste duty of care.
UniclassPrCodes	The primary data within the class is the Uniclass Pr codes and their corresponding names adopted from the NBS platform.	The Pr codes in the class serve as a primary variable to filter materials by mapping the data from the BIM model with ontology data.
Unit	This class represents the declaration units of all technical and environmental information. It includes the unit of density, dimensional (length, breadth, and thickness) unit, service life (calendrical), GWP metrics based on the EPD declaration unit, and unit of WGR based on GFA.	It is one of the core classes added to the ontology to ensure consistency of the waste results.

Source: Drafted by the author

5.2.4.3 EPDProductType Class Hierarchy

Figure 5.4 depicts the inferred class hierarchy of the *EPDProductType* class. This class lists the building products that have manufacturer-specific EPD certificates. Since the framework is designed to work with design stage information that meets the requirements of LOMD 3 (discussed in Section 3.2.5), applying product-specific data extracted from the EPDs is appropriate and ensures the accuracy and relevance of the results.

There are two types of classes called defined and primitive in the Description Logic language. The former has both necessary (super-class) and sufficient conditions (equivalent class), while the latter has only the necessary conditions (super-class) for a membership. The orange coloured are defined classes, the yellow coloured are primitive classes, and the arrow denotes that a specific class is a sub-class of another class (rdfs:subClassOf). For example, the class *EPD_ConcreteBlocks* is equivalent to another class expression, *EPD_ConcreteMasonryUnits*. This states that the two classes are alternative names for the same set of individuals so they would have the same individuals in any interpretation. Figure 5.5 shows the instance of the *EPD_ConcreteBlocks* class. The *EPD_ConcreteBlocks* is

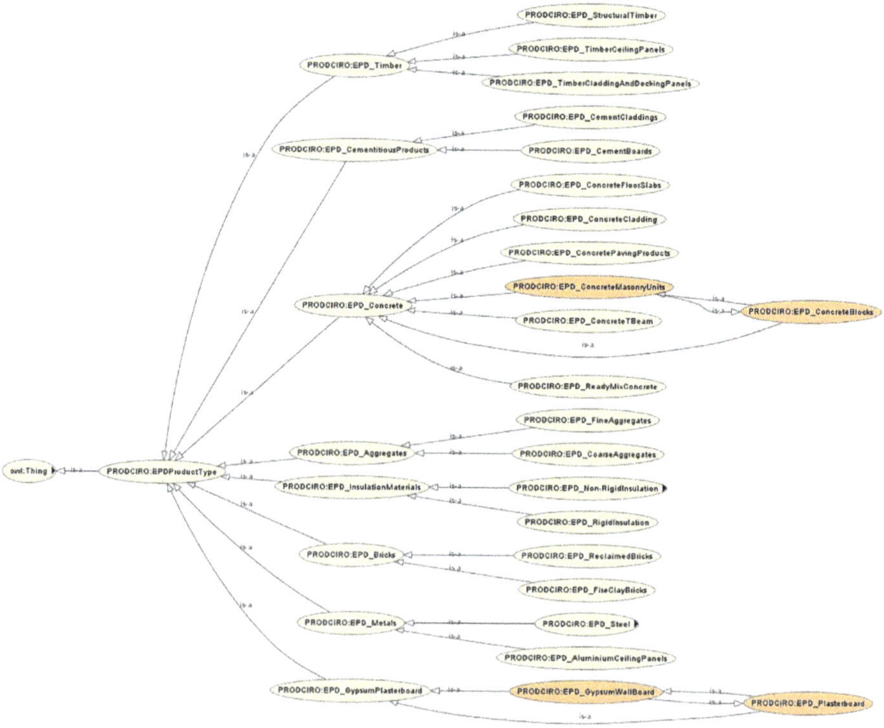

Figure 5.4 Inferred Class Hierarchy of EPDProductType – a Sample

Source: Drafted by the author

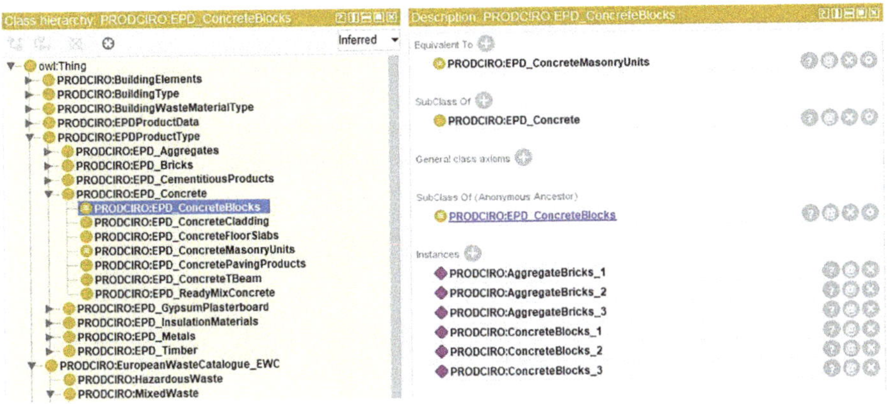

Figure 5.5 Asserted Instances of EPD_ConcreteBlocks

Source: Drafted by the author

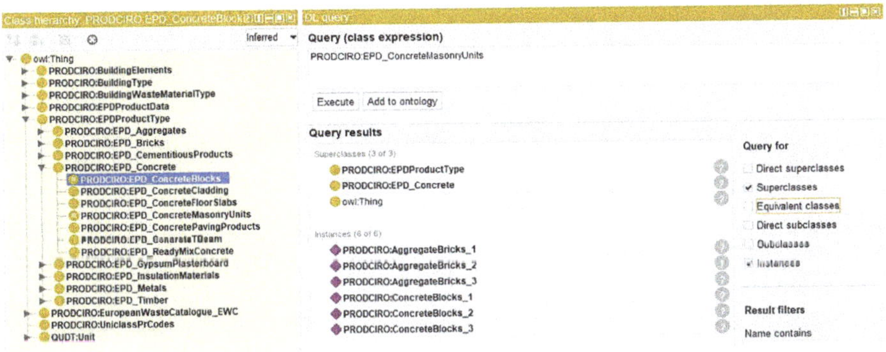

Figure 5.6 Inferred Instances of EPD_ConcreteMasonryUnits

Source: Drafted by the author

equivalent to *EPD_ConcreteMasonryUnits*, so the instances of both classes are similar (Figure 5.6). The same applies to *EPD_Plasterboard* and *EPD_GypsumWallBoard* under the *EPD_GypsumPlasterboard* class. Both classes' inherent definitions and material properties are similar, but only the name varies with geographical location. So, they are defined to act as equivalent to each other.

The critical aspect is that although the selected middle-out approach gives the flexibility to develop a class hierarchy, the design aligns with the following rule: 'If class A is a sub-class of class B, then every instance of class A will be inferred as an instance of class B' [2]. So, if *ConcreteBlocks_1* is an instance of *EPD_ConcreteBlocks*, then it is also an instance of middle-class *EPD_Concrete* (Figure 5.7) and the super-class *EPDProductType* (Figure 5.8).

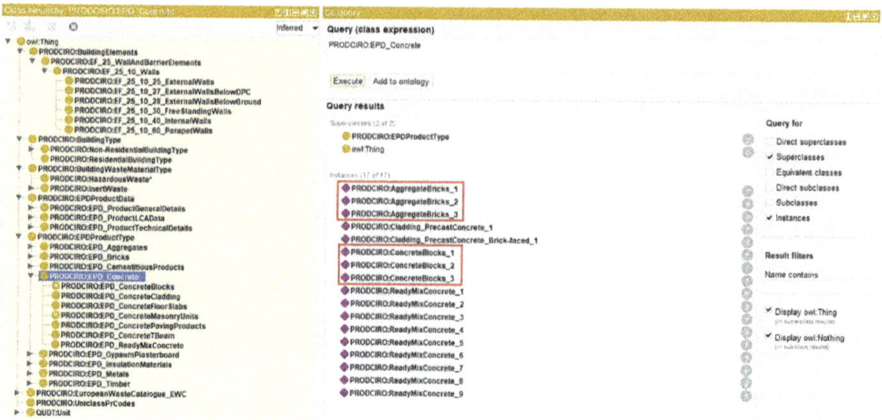

Figure 5.7 Inferred Instances of EPD_Concrete

Source: Drafted by the author

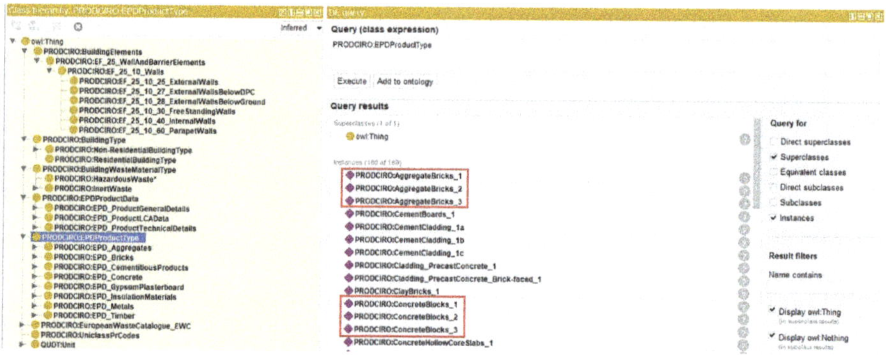

Figure 5.8 Inferred Instances of EPDProductType

Source: Drafted by the author

5.2.4.4 EPDProductData Class Hierarchy

Data from EPDs is one of the primary knowledge sources for ontology modelling. EPDs provide generic, technical, GWP, and WM routes for the building products. The structure of the *EPDProductData* class represents the structure of EPD certificates (Figure 5.9). It is important to note that the digital LCA data format is not preferred due to limited generic and technical details.

Some generic and technical data collected for modelling include product description, application, density, mass, dimensions (length, breadth, and thickness), service life, manufacturer, and manufacturing location (identified from Section 4.5.2). This information is related to the *EPDProductData* class via the middle-level classes *EPD_ProductGeneralDetails* and *EPD_TechnicalDetails*.

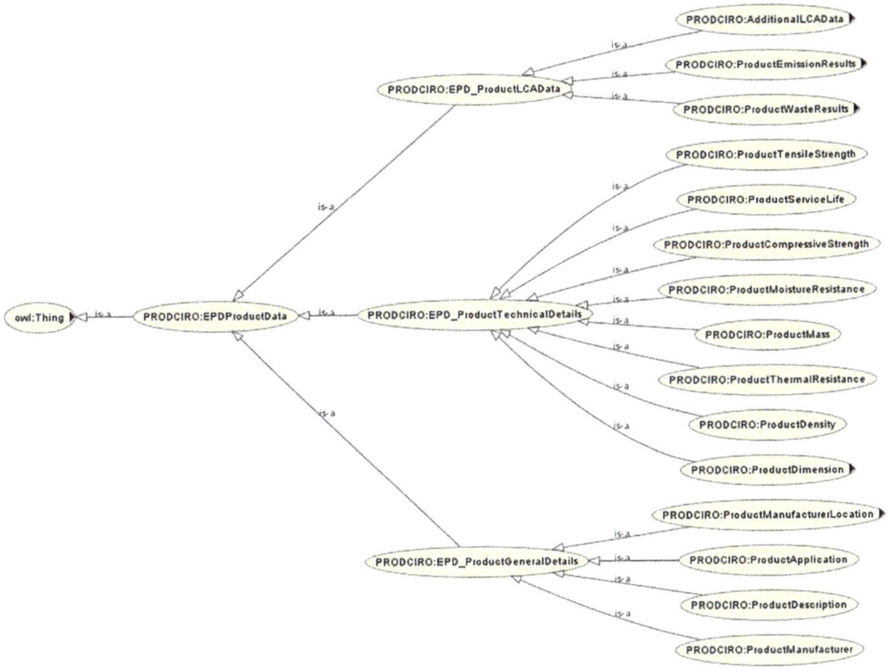

Figure 5.9 Inferred Class Hierarchy of EPDProductData – a Sample

Source: Drafted by the author

The *EPD_ProductLCAData* class consist of environmental information, including waste generation, GWP levels, EoL scenarios, and use of secondary materials during product manufacturing. The middle-level classes, including *ProductWasteResults* and *ProductEmissionResults*, hold this information. In addition to these details, the classification of each product type is supported using the *UniclassPrCodes* class, which contains information that helps the user map the ontology knowledge appropriately with the BIM material data without any inconsistencies. A description of how the information is mapped is discussed in Section 6.3.3 – Step 4.

5.2.4.5 BuildingWasteMaterialType Class Hierarchy

The second most crucial data type for PRODCIRO development is the WGR from BRESmartWaste reports, defined as the *BuildingWasteMaterialType* class. Thus far, CWQ models have adopted limited methodologies to classify waste, which is one of the significant gaps concerning its applications at a broader scale. In some studies, CW is classified based on its occurrence as avoidable and non-avoidable waste. In contrast, the reactive nature of waste accounts for its classification as inert, non-inert, and hazardous waste. However, the higher the granularity of waste classification, the higher the chances for waste reduction [19]. Therefore, a detailed classification approach aligning with the EWC classification system is implemented in the ontology modelling.

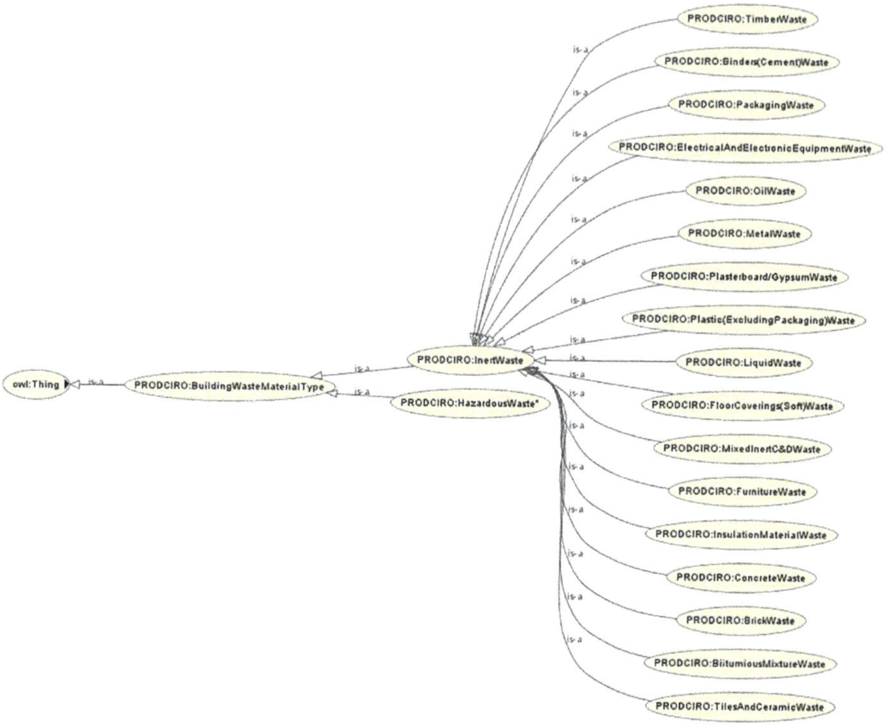

Figure 5.10 Inferred Class Hierarchy of BuildingWasteMaterialType
Source: Drafted by the author

Figure 5.10 shows the sub-classes of *BuildingWasteMaterialType*. The waste materials are initially classified based on their reactive nature as *InertWaste and HazardousWaste**. Furthermore, the material composition of waste is added to the classification to ensure the granularity of CWQ results. This type of waste classification will have three significant benefits for the users: (i) quantifying the total waste from the super-class, (ii) determining the reactive nature of waste for deciding its disposal route from the middle-level classes, and (iii) maximise the reuse and upcycling potentials of waste materials from low-level classes. Besides, each material type is allocated with an EWC code to ease waste reporting at EoL. The EWC codes are added to the *EuropeanWasteCatalogue_EWC* class. Incorporating these codes facilitates the CWQ model to adopt a systematic CW reporting that aligns with the legal duty of care.

5.2.4.6 *BuildingType Class Hierarchy*

Besides the granularity associated with CW classification types, the literature review emphasised that the relationship between the project features and the WGRs should be clearly defined in the quantification process. This is because the WGR

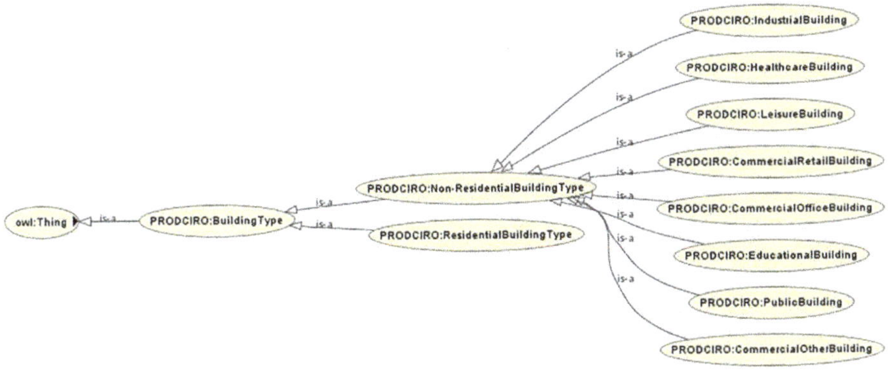

Figure 5.11 Inferred Class Hierarchy of BuildingType

Source: Drafted by the author

is influenced by several project features, including building usage, geographical location, architectural type, procurement method, construction stages, construction type, and other temporal scenarios (discussed in Section 2.7.2). Hence, these factors are taken into account while developing the ontology. Despite the impact of all these factors, the BRESmartWaste reports, the knowledge source selected for the research (discussed in Section 5.2.1), explicitly show that building usage type impacts the WGR of materials during the construction and installation stages. Therefore, the influence of building usage type on WGRs of materials is demonstrated via the *BuildingType* class.

Figure 5.11 presents the middle-level and low-level classes of the top-class *BuildingType*. The first classification is broader and has two classes, *ResidentialBuildingType* and *Non-ResidentialBuildingType*. The *ResidentialBuildingType* class has no further classification. Besides, the *Non-ResidentialBuildingType* class is further classified as *CommercialOfficeBuilding, CommercialRetailBuilding, CommercialOtherBuilding, PublicBuilding, EducationalBuilding, IndustrialBuilding, LeisureBuilding,* and *HealthcareBuilding.* The building type classification will differ depending on the approach adopted. However, the classification used in BRESmartWaste reports was adopted to maintain consistency in the results in PRODCIRO.

In addition to the building usage, the geographical location and construction type are implicitly incorporated into the analysis. This is because the CW data from the BRESmartWaste reports are specific to the UK and in-situ construction type. Overall, accounting for the building usage type, construction type, and geographical location as modelling variables ensures the accuracy and relevance of the CWQ results.

5.2.4.7 BuildingElements Class Hierarchy

As discussed in Section 4.5.2, the information on building materials' application is considered vital among industry professionals to make informed decisions.

Therefore, the application details of each building material are extracted from the EPD documents. Since there is no systematic approach to adding applications in the EPDs, the applications may be provided at the building or component level. So, the research classified materials based on their application at the component level, which is demonstrated using the *BuildingElements* class. Figure 5.12 shows *BuildingElements* class hierarchy. Since the application of the framework via an ontology is at the proof of concept level, only one element type *(EF_25_10_Walls)* is selected to demonstrate the workflow.

It can be seen that low-level classes, including *EF_25_10_25_ExternalWalls, EF_25_10_40_InternalWalls,* and *EF_25_10_60_ParapetWalls,* are used to classify the wall elements. So, an instance of class *EF_25_10_25_ExternalWalls* is also an instance of class*es EF_25_10_Walls, EF_25_WallAndBarrierElements* and the super-class *BuildingElements,* aligning with the ontology rules. This allows the framework to inform optimal material choices based on their applications.

5.2.4.8 Unit Class Hierarchy

Unit conversion is a significant factor affecting the accuracy and consistency of waste results. So, eight major types of units supporting CWQ are added to the PRODCIRO. It includes *DimensionUnit* (metric system), *AreaDensityUnit* (kg/m^2), *DensityUnit* (kg/m^3), *WGRUnit* of materials (kg/m^2), *WasteAndResource Unit* (kg), *GWPUnit* (kgCO2e), *ServiceLifeUnit* (calendrical), *and EPDDeclarationUnit* (as per EPD data). Figure 5.13 presents the unit types formalised in the ontology. All the described sub-classes are merged under the super-class *QUDT:Unit.*

5.2.5 Define the Properties of Classes – Slots

The properties define the relationships of classes. The three main property types are object, data, and annotation properties. The following section discusses each property type and its definitions in PRODCIRO.

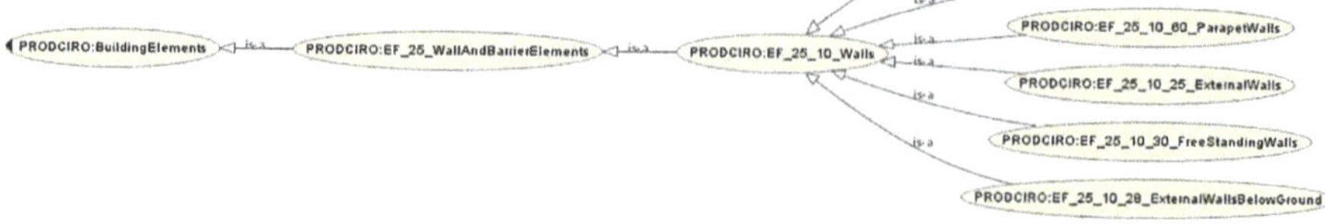

Figure 5.12 Inferred Class Hierarchy of BuildingElements

Source: Drafted by the author

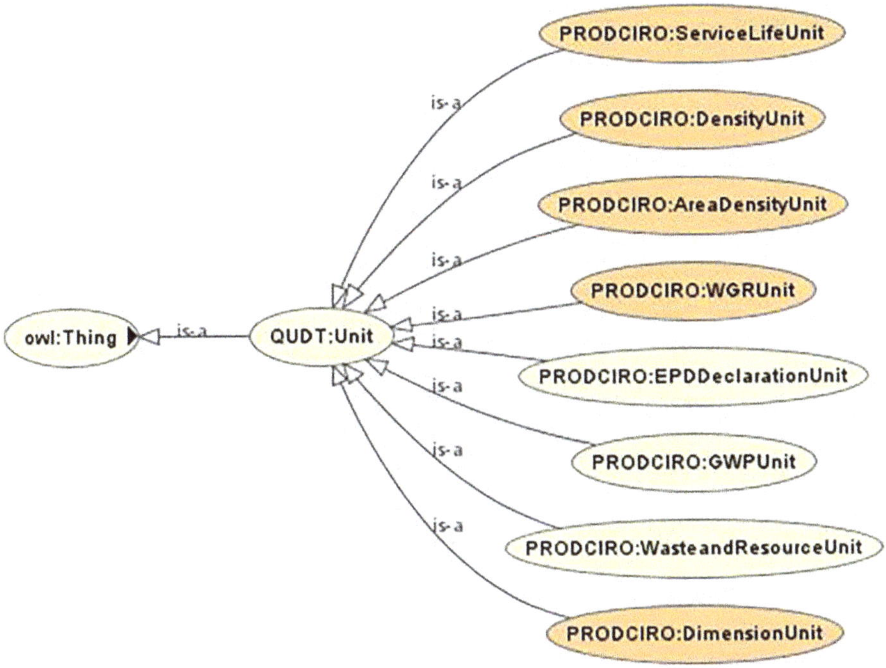

Figure 5.13 Inferred Class Hierarchy of Unit

Source: Drafted by the author

5.2.5.1 Object Properties

Object properties are the predicate(s) that connect the subject(s) with an object(s), which are called triples in the RDF representation (discussed in Section 3.4.5). The subject(s) and object(s) are individuals of two different classes, where an object of one triple could be a subject of another triple. To cite an example, Figure 5.14 shows that the *EPDProductType* and *ProductManufacturer* classes form a triple via the object property *isManufacturedBy*. Here, the instance of *EPDProductType* is the subject, the instance of *ProductManufacturer* is the object, and *isManufacturedBy* is the predicate. Besides, the instances of the same classes are connected via another object property, *manufactures*. In this case, the instance of *ProductManufacturer* is the subject, the instance of *EPDProductType* is the object and *manufactures* is the predicate. So, the subject of one triple can be an object of another triple. The subject and object are called domain and range, respectively, in the RDFS description. Table 5.3 shows the complete list of object properties added to PRODCIRO and a description of each object property with their domain(s) and range(s).

In addition, RDF/OWL has additional characteristics of property axioms like RDFS schema, relational, global cardinality, and logical property constraints [20].

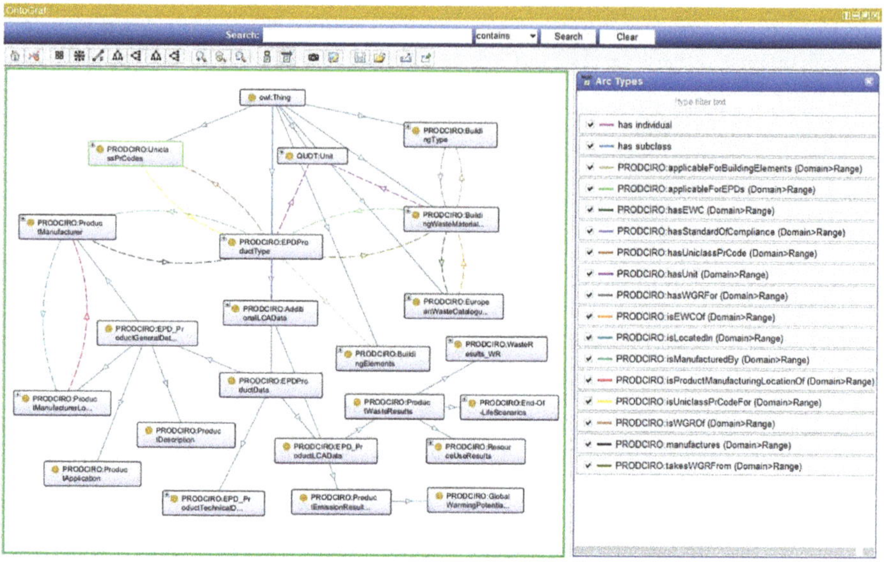

Figure 5.14 PRODCIRO Object Properties With Domain(s) and Range(s)

Source: Drafted by the author

The rdfs:domain and rdfs:range are RDFS schema constructs that could be defined as an object property. It asserts that the subject(s) and the object(s) of a property should belong to a specific class(es). Besides, the subject(s) and object(s) of all the listed object properties are asserted to class description(s) in the ontology. For example, the subject and object of a property *hasEWC* are asserted to the class *BuildingWasteMaterialType* and *EuroepeanWasteCatalogue_EWC,* respectively. Although there is no defined rule for having a domain and range, the object properties would be either unidirectional or bidirectional based on the domain-range relations. The unidirectional object properties have no link with other object properties. In contrast, the bidirectional connection shows that a specific object property is an equivalent or inverse of another. To cite an example, *isManufacturedBy* is an inverse property of *manufactures* (Figure 5.14 and Table 5.3). It means that if a product *isManufacturedBy* a specific manufacturer, then the specific manufacturer *manufactures* that product. This bidirectional connection enables the ontology to infer the queries and make meaningful interpretations logically.

Furthermore, functional properties are adopted in PRODCIRO. These characteristics define object properties as having only one unique value for each instance and no two distinct values. For example, the object property *isManufacturedBy* is functional because an EPD is specific to a manufacturer, so an individual defined under the class *EPDProductType* can have only one manufacturer.

Table 5.3 List of PRODICRO Object Properties, Description, and Their Domain(s) and Range(s)

Object property	*Description*	*Domain(s)*	*Range(s)*
applicableForBuildingElements	Defines the relationship between the classes of *EPDProductType* and *BuildingElements*.	*EPDProductType*	*BuildingElements*
applicableForEPDs	Defines the relationship between the classes of *BuildingWasteMaterialType* and *EPDProductType*. It is also an inverse property of *takesWGRFrom*.	*BuildingWasteMaterialType*	*EPDProductType*
hasEWC	Defines the relationship between the class *BuildingWasteMaterialType* and *EuroepeanWasteCatalogue_EWC*. It is also an inverse property of *isEWCOf*.	*BuildingWasteMaterialType*	*EuroepeanWasteCatalogue_EWC*
hasUniclassPrCode	Defines the relationship between the classes of *EPDProductType* and *UniclassPrCodes*. It is also an inverse property of *isUniclassPrCodeFor*.	*EPDProductType*	*UniclassPrCodes*
hasWGRFor	Defines the relationship between the classes of *BuildingType* and *BuildingWasteMaterialType*. It is also an inverse property of *isWGROf*.	*BuildingType*	*BuildingWasteMaterialType*
isEWCOf	Defines the relationship between the class *EuroepeanWasteCatalogue_EWC* and *BuildingWasteMaterialType*. It is also an inverse property of *hasEWC*.	*EuroepeanWasteCatalogue_EWC*	*BuildingWasteMaterialType*
isLocatedIn	Defines the relationship between the class *ProductManufacturer* and *ProductManufacturingLocation*. It is also an inverse property of *hasEWC*. It is also an inverse property of *isProductManufacturingLocationOf*.	*ProductManufacturer*	*ProductManufacturingLocation*
isManufacturedBy	Defines the relationship between the classes of EPDProductType and *ProductManufacturer*. It is also an inverse property of *manufactures*.	*EPDProductType*	*ProductManufacturer*
isProductManufacturingLocationOf	Defines the relationship between the class *ProductManufacturingLocation* and *ProductManufacturer*. It is also an inverse property of *isLocatedIn*.	*ProductManufacturingLocation*	*ProductManufacturer*
isUniclassCodePrFor	Defines the relationship between the classes of *UniclassPrCodes* and *EPDProductType*. It is also an inverse property of *hasUniclassPrCode*.	*UniclassPrCodes*	*EPDProductType*
isWGROf	Defines the relationship between the classes of *BuildingWasteMaterialType* and *BuildingType*. It is also an inverse property of *hasWGRFor*.	*BuildingWasteMaterialType*	*BuildingType*
manufactures	Defines the relationship between the classes of *ProductManufacturer* and *EPDProductType*. It is also an inverse property of *isManufacturedBy*.	*ProductManufacturer*	*EPDProductType*
takesWGRFrom	Defines the relationship between the classes of *EPDProductType* and *BuildingWasteMaterialType*. It is also an inverse property of *applicableForEPDs*.	*EPDProductType*	*BuildingWasteMaterialType*
hasUnit	Defines the relationship between the class 1) *EPDProductType* and *Unit* 2) *BuildingWasteMaterialType* and *Unit*	*EPDProductType*, *BuildingWasteMaterialType*	*Unit*

Source: Drafted by the author

5.2.5.1.1 RELATIONSHIP BETWEEN THE CLASSES – *EPDPRODUCTTYPE, PRODUCTMANUFACTURER*, AND *PRODUCTMANUFACTURINGLOCATION*

The following section will discuss the relationship established within the four object properties: *isManufacturedBy, manufactures, isLocatedIn, and isProductManufacturingLocationOf.* Based on the analysis of primary data collected via interviews, manufacturers and manufacturing units seem to have a significant influence on waste and carbon levels (Section 5.3). Hence, PRODCIRO establishes the relationship between the individuals of product class *EPDProductType* and its manufacturing details.

Figure 5.15 shows that the *EPD_Plasterboard* is an equivalent class of *EPD_GypsumWallBoard,* a sub-class of EPDProductType, and has eight individuals. The figure shows that the manufacturers of the first six plasterboard types (*Plasterboards_1* to *Plasterboards_6*) are the same, while others (*Plasterboards_7* and *Plasterboards_*8) have different manufacturers. Each manufacturer has one or more manufacturing units across different geographical locations. To cite an example, Figure 5.15 shows that the product manufacturer, 'Saint_Gobain', has four manufacturing units across the UK. Therefore, the object properties connecting *EPD_Plasterboard, ProductManufacturer*, and *ProductManufacturingLocation* classes allow users to optimise the most appropriate location by calculating the distance between the construction site and manufacturing units. Furthermore, this would potentially reduce transportation distance and long-time storage of materials, causing waste and its related impacts, one of the CW indicators.

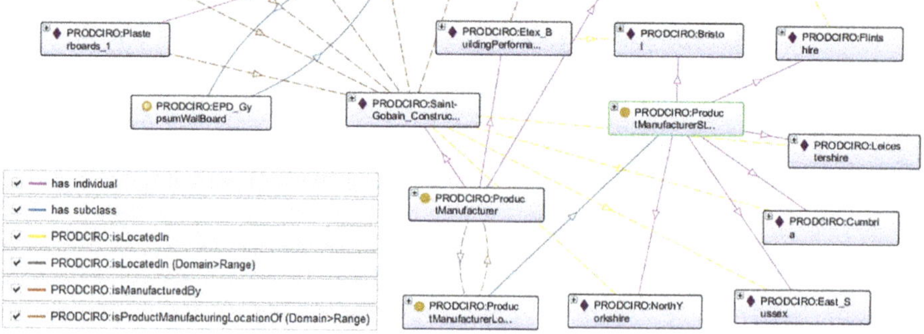

Figure 5.15 Object Properties Connecting EPD_Plasterboard, ProductManufacturer, and ProductManufacturingLocation

Source: Drafted by the author

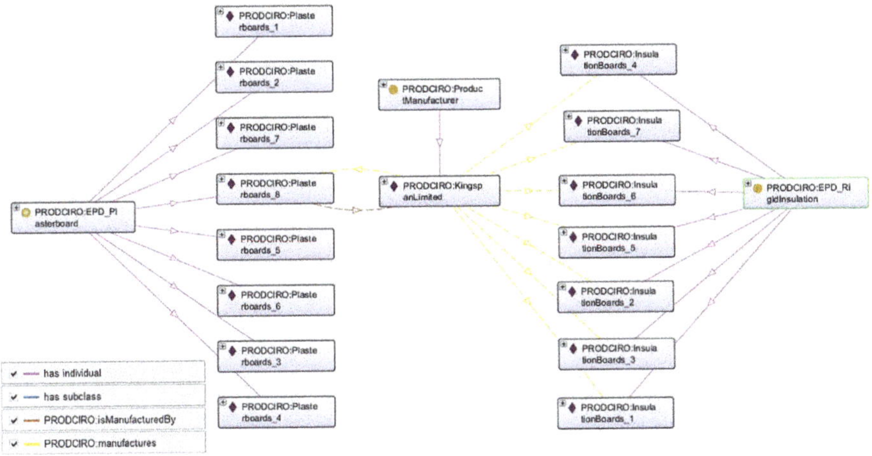

Figure 5.16 Object Properties Connecting the instances of EPD_Plasterboard, EPD_Rigid-Insulation, and ProductManufacturer Classes

Source: Drafted by the author

Besides, in some cases, a specific manufacturer might produce multiple products. For example, Figure 5.16 shows that the manufacturer of *Plasterboards*_8 also manufactures a few types of insulation materials (*InsulationBoards_1* to *InsulationBoards_7*). Hence, selecting these manufacturers could significantly minimise material diversity and ease decision-making. Overall, the output of PRODCIRO will provide all the necessary information deemed significant for CW-informed decision-making.

5.2.5.1.2 RELATIONSHIP BETWEEN THE CLASSES – *EPDPRODUCTTYPE* AND *BUILDINGELEMENTS*

Since the application of building products is one of the principal CW indicators (discussed in Section 4.5.2), they are included in PRODCIRO. For example, comparing individuals *Plasterboards_1* and *Plasterboards_2*, the former applies only to internal walls (*EF_25_10_40_InternalWalls*), while the latter extends its application to internal (*EF_25_10_40_InternalWalls)* and external walls (*EF_25_10_25_ExternalWalls),* as shown in Figure 5.17. Among the two individuals, the latter has more chances of reaching the top rank during optimisation due to minimised waste generation resulting from less diversity of material types, given the minimal environmental impacts. This testing is facilitated by design and material information within the BIM model.

To illustrate, if the selected plasterboard of a BIM element is an individual of the class *EF_25_10_25_ExternalWalls*, the users will be prompted with the product list based on a Boolean condition, individuals of the class *EPD_Plasterboard* AND individuals related to the class *EF_25_10_25_ExternalWalls* via

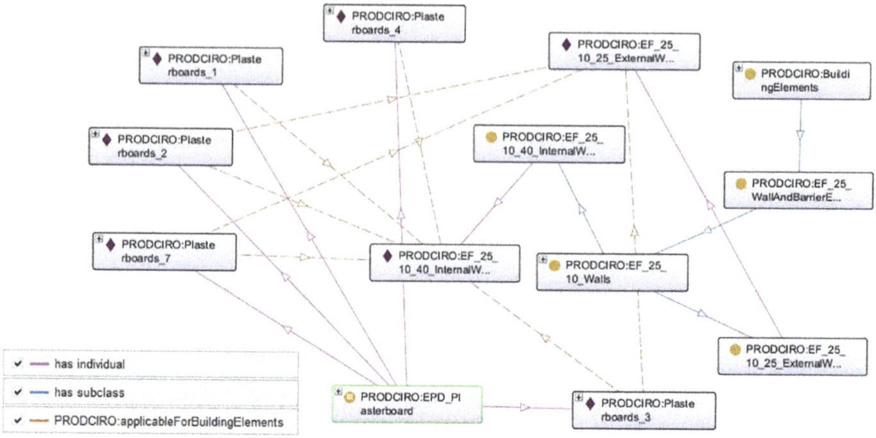

Figure 5.17 Object Properties Connecting Instances of EPD_Plasterboard and BuildingElements Classes

Source: Drafted by the author

applicableForBuildingElements property. Therefore, integrating the BIM data into the analysis demonstrates the relationship between the classes *EPDProductType* and *BuildingElements*. However, it is important to note that not all EPDs provide the application of their products at the same level of detail. So, accounting for the limitations and mapping them appropriately is vital. This limitation is managed in this research by using the Uniclass classification system and our understanding of the building materials and their application in the built environment.

5.2.5.1.3 RELATIONSHIP BETWEEN THE CLASSES – *EPDPRODUCTTYPE, BUILDINGWASTEMATERIALTYPE, EUROPEANWASTECATALOGUE_EWC*, AND *BUILDINGTYPE*

The waste data in the EPDs are presented as 'non-hazardous waste disposed' and 'hazardous waste disposed' for the specific life-cycle stages within the defined system boundary. Also, the WGR during construction is estimated to be an average of 5%–20%, depending on the product types in the EPDs. However, the literature review proves that incorporating average WGRs leads to uncertainty in the results (discussed in Section 2.7.2). So, the differences based on its associated variables should be established for accurate quantification. Based on the statistical analysis of several waste-influencing variables, BRESmartWaste data showed that building usage is one of the variables that significantly impacts the WGR of materials. Therefore, accounting for these impacts, the waste data from BRESmartWaste reports and EPDs are used for quantification in the construction/installation and manufacturing stages, respectively. The impacts of CW generated during the operation and maintenance stages are estimated using the lifetime analysis approach due to inadequate databases. A detailed description of the quantification approaches for each life cycle stage is provided in Section 6.3.3 – Step 5.

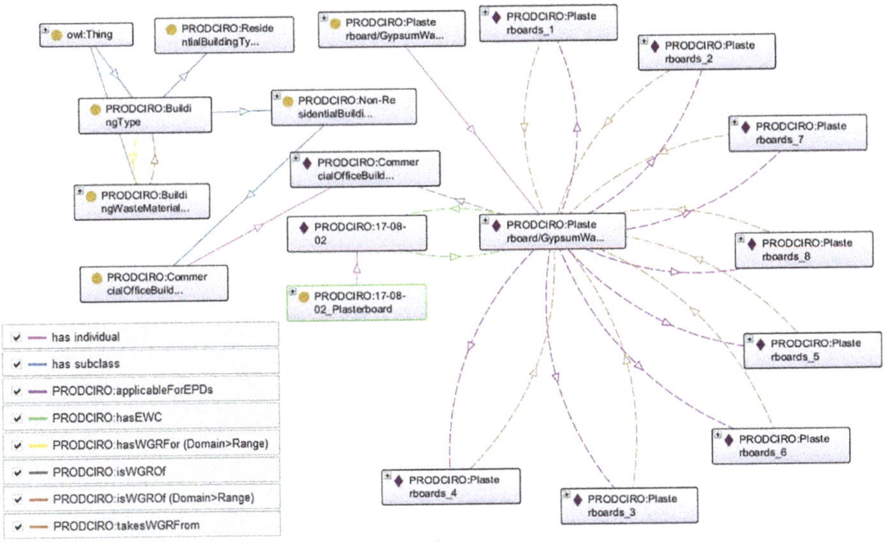

Figure 5.18 Object Properties Connecting EPD_Plasterboard, Plasterboard/GypsumWaste, BuildingType, and EuropeanWasteCatalogue_EWC

Source: Drafted by the author

However, the EPD data are product-based, and WGRs are material-based, causing discrepancies in using them directly in the quantification process. So, a relationship between them is established via object properties *applicableForEPDs*, *takesWGRFrom, hasEWC*, and *isEWCOf*. Figure 5.18 shows the relationships of classes *EPD_Plasterboard, Plasterboard/GypsumWaste*, and EWC *17–08–02_Plasterboard*. It can be seen from the figure that the waste type *Plasterboard/GypsumWaste_1* applies to all individuals of class *EPD_Plasterboard*. This is due to the similarities expressed via the individuals of EWCs, which is *17–08–02* in this case. Besides, the *Plasterboard/GypsumWaste_1* individual is related to *CommercialOffficeBuilding_1* using the *isWGROf* object property. As seen from the graph (Figure 5.18), these two individuals are asserted to *BuildingWasteMaterialType* and *BuildingType* classes, respectively. The relationship between instances of these two classes shows that the WGR of each material depends on the building usage type. Hence, integrating WGR based on building usage type and EPD results could enhance the accuracy and relevance of quantification results. Furthermore, EWCs would enhance waste classification and reporting by aligning with legal standards.

5.2.5.1.4 RELATIONSHIP BETWEEN THE CLASSES – *EPDPRODUCTTYPE* AND *UNICLASSPRCODES*

Since information standardisation and BIM interoperability are at the forefront of ontology development, the Uniclass Pr codes are incorporated in PRODCIRO. For example, the *EPD_ConcreteBlocks*, which is an equivalent class of

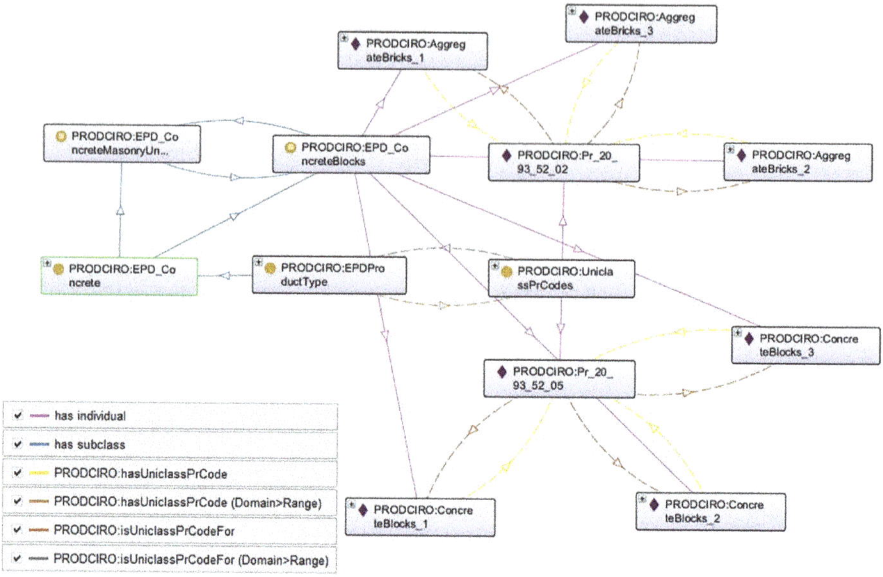

Figure 5.19 Object Properties Connecting EPDProductType and UniclassPrCodes

Source: Drafted by the author

EPD_ConcreteMasonryUnits and a sub-class of *EPD_Concrete,* is related to Uniclass codes via the object properties *hasUniclassPrCode* and *isUniclassPrCodeFor* (Figure 5.19).

Besides, the Uniclass Pr codes have three and four-tier classifications. The fourth tier, the extension of third tier codes, is based on the detailed material composition of a specific product, providing a high level of certainty during optimisation. To cite an example, 'Pr_20_93_52' is classified as 'Masonry Walling Units', which is further classified based on its material composition as 'Clay bricks (Pr_20_93_52_15)', 'Autoclaved aerated concrete blocks (Pr_20_93_52_05)', and 'Aggregate concrete bricks (Pr_20_93_52_02)'. In PRODCIRO, the four-tier Uniclass Pr codes are added to the classes to enhance the granularity and accuracy of material classification.

5.2.5.2 *Data Properties*

Data properties are predicates that help to create a relationship between an individual and an attribute data. Like object properties, data property axioms also have additional RDFS schema constructs rdfs:domain and rdfs:range. In this case, the domain asserts that the subject(s) of the property belongs to a specific class expression. In contrast, the range asserts that the property's value must belong to a data value in a specific data range. The data range in RDF is borrowed from XML schema data types, and it would be a primitive or enumerated data type used to

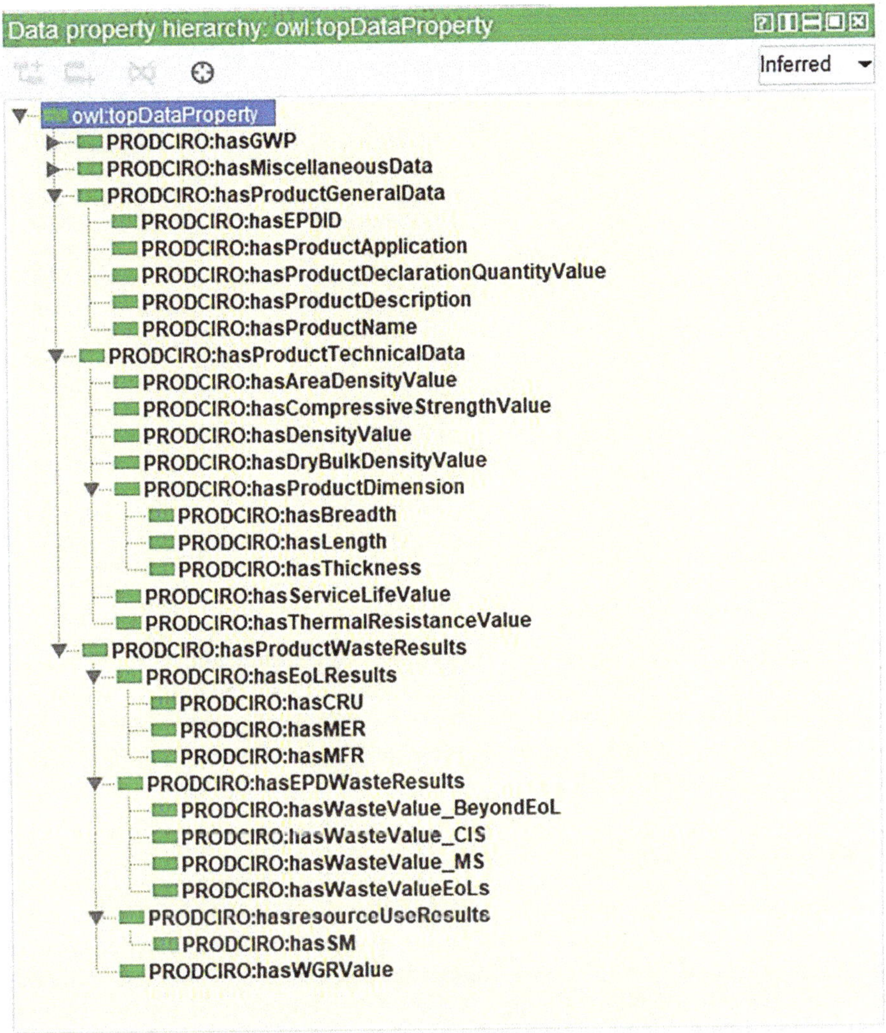

Figure 5.20 List of Data Properties in PRODCIRO

Source: Drafted by the author

restrict the range of properties [20]. Figure 5.20 shows an extract of data properties of PRODCIRO.

Table 5.4 shows the data properties of PRODCIRO supporting CWQ, along with its domain(s) and range(s). These data properties are established based on the primary and secondary information collected from the knowledge sources listed in Section 5.2.1. Each data property significantly impacts the granularity of waste and carbon results. For example, the data properties *hasProductDeclarationQuantityValue* and *hasDensityValue* within the properties of *hasProductGeneralData* and *hasProductTechnicalData* allow the user to estimate the mass of the product using the domain specification. So, real-life test-case-building data is used to evaluate the accuracy and adaptability of the semantic framework demonstrated via an ontology and BIM-integrated UI. The tool will be tested using the workflow discussed in the previous section. A detailed discussion of the test case and operational and quantification results is provided in this section.

6.4.1 Test Case Description

The test case selected for validation is a two-storey educational building located in the UK. The GFA of the building estimated from the BIM model is 1860 m^2. The obtained BIM model was at the design stage with an LOMD 3. Although the increase in project stages increases the LOMDs and provides detailed material information, the chances for optimisation are highly compromised at later stages. Hence, the tool is designed to support designers and architects in exploring alternative material choices from the early design stages (discussed in Section 4.6). The BIM model of the case-study building utilised the Common Arrangement of Work Sections (CAWS) codes to classify the elements and materials applied in the design. However, one of the prerequisites to ensure accuracy is to have the Uniclass codes for model elements to connect BIM data with the semantic model (discussed in Section 6.3.1). Therefore, an additional relationship is manually established in the analysis to convert the CAWS to the Uniclass codes. The NBS BIM object library facilitated this conversion.

6.4.2 Test Case Operational Results

This section will provide the operational results of the test case. These results are the outputs of the component, 'Perform Waste Analysis' (discussed in Section 6.3.3). This validates the first check: ensuring that the proposed framework performs based on the defined requirements and undergoes the complete process seamlessly.

6.4.2.1 Element Selection

Following the creation of the custom parameters, the user is prompted to provide basic information about the BIM design (Figure 6.15) and select an element for waste analysis (Figure 6.16).

6.4.2.2 Existing Material Layers of the Selected Element

The selected test case building has multiple structural elements, including walls, floors, roofs, and ceilings. Each element is comprised of either a single material or multiple material layers. Multiple elements sharing the same features make up an element family, and multiple families make up a category. The element selected to test the operation of the framework is 'EW-01 Brickwork with SFS'. This element is an 'external wall element' that falls under the 'Basic Wall' family of type 'EW-01 Brickwork with SFS'. This wall family is a compound structure comprising seven different material layers. The material layers include bricks as an outer

Table 5.4 Data Properties of PRODICRO Supporting CWQ

Data property	*Domain*	*Range*
hasWGRValue	*BuildingWasteMaterialType*	xsd:float
hasProductDeclarationQuantityValue	*EPDProductType*	xsd:integer
hasEPDID	*EPDProductType*	xsd:string
hasProductName	*EPDProductType*	xsd:string
hasLength	*EPDProductType*	xsd:float
hasBreadth	*EPDProductType*	xsd:float
hasThickness	*EPDProductType*	xsd:float
hasDensityValue	*EPDProductType*	xsd:float
hasAreaDensityValue	*EPDProductType*	xsd:float
hasServiceLifeValue	*EPDProductType*	xsd:integer
hasDryBulkDensityValue	*EPDProductType*	xsd:float
hasGWPValue_A1-A3	*EPDProductType*	xsd:float
hasWasteValue_MS	*EPDProductType*	xsd:float
hasCRU	*EPDProductType*	xsd:float
hasMFR	*EPDProductType*	xsd:float
hasMER	*EPDProductType*	xsd:float
hasSM	*EPDProductType*	xsd:float

Source: Drafted by the author

the simple mathematical expression: the volume of the product times the density of the product gives the total mass. This, in turn, provides the quantity of waste and embodied carbon of waste materials. Since the calculations are expressed using SPARQL, Section 5.4 will show the application of each property in the CWQ process using a sample SPARQL query.

5.2.5.3 *Annotation Properties*

The third property type, annotation property, is a way to label, describe, or comment on any ontology entities with URI using annotations. The annotations briefly describe entities, including classes, object properties, data properties, individuals, and even the complete ontology. Figure 5.21 shows the annotation properties of an individual *Plasterboards_1*. The annotation rdfs:label represents the

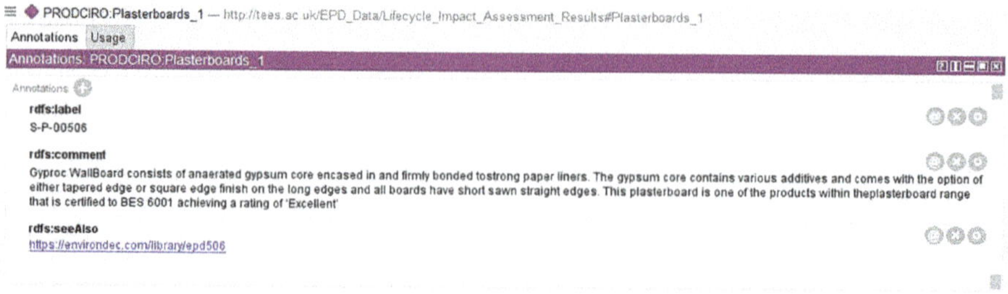

Figure 5.21 Annotation Properties of Plasterboards_1

Source: Drafted by the author

human-readable version of the entity, rdfs:comment provides a human-readable description about the product and rdfs:seeAlso provides a URL to connect to the actual data source (here it is an EPD document).

5.2.6 Define the Facets of the Slots

The facets are the constructs established as domain, range, and data type accepted. Noy and McGuinness [2] classify the facets as instance-type and value-type slots. The former represents the domain(s) and range(s) of the class expressions, while the latter represents the data-type restriction(s) given to a property. The facet types included in PRODCIRO development are explained in 5.2.5.

Figure 5.22 depicts an example of instance-type slots given to *isProductManufacturingLocationOf* object property. Here, a restriction is imposed that the subject and object of the property should belong to the class *ProductManufacturingLocation* and *ProductManufacturer*, respectively.

Figure 5.23 depicts an example of value restrictions and instance-type slots given to the data property *hasWGRValue*. Here, the data property imposes an instance-type slot that accepts only individuals relating to the *BuildingWasteMaterialType* class. It also imposes a value-type restriction that it should be a float number (xsd:float) to perform a quantitative CW assessment.

5.2.7 Create Instances

The instances are the basic ground-level entities represented by a class. In other words, instances sharing common facts are conceptualised as a class. They are created based on how granular the information in the semantic model needs to be. The granularity depends on ontology applications [2]. The waste data of each material is based on its building usage, and the technical and carbon impact values

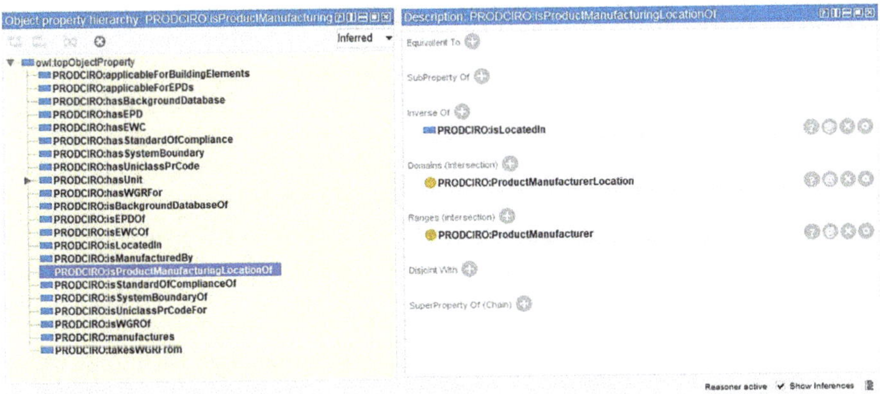

Figure 5.22 Facets of isProductManufacturingLocationOf Object Property

Source: Drafted by the author

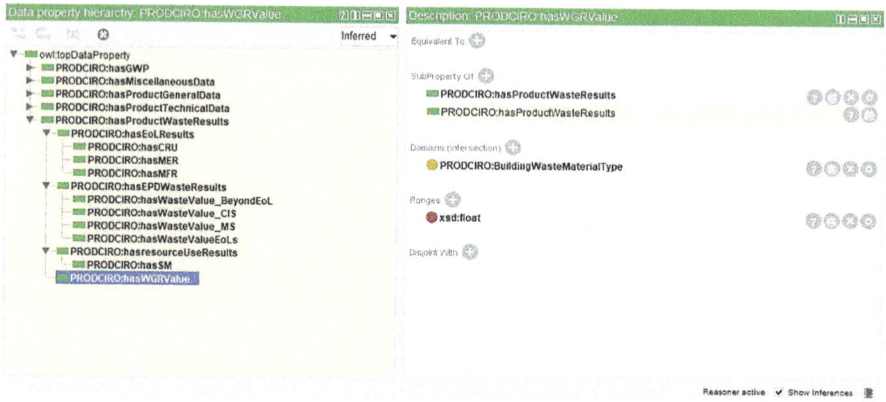

Figure 5.23 Facets of hasWGRValue Data Property

Source: Drafted by the author

of materials are based on the EPD reports of different manufacturers. These two entities benchmark the construction of instances in PRODCIRO. For example, *EPD_Bricks* is a sub-class of *EPDProductType*. If multiple manufacturers manufacture bricks, there might be multiple EPDs (instances) for brick material. So, the best possible brick type (instance) will be filtered from the various material choices (clustered under a class), aligning with the optimisation conditions. The instances created in the ontology hold generic, technical, waste, and carbon data for materials. The individuals of each class decide the waste generation and its impact levels to facilitate optimisation. Since the transformative rules and conditions should be designed individually for each element, walls are considered the primary instance in PRODCIRO. So, the basic structural materials, including bricks, concrete blocks, steel, timber, plasterboard, and insulation materials supporting the wall designs, are included within the scope of the research. These materials act as instances in PRODCIRO.

The technical and environmental impact data of materials are captured from EPD documents. Despite the availability of product-specific data in the EPDs, a few aspects require consideration. Firstly, EPDs are comparable only if they share the same product category rules (PCR). The European Standard EN15804 is a global standard that provides PCR for declaring EPDs for construction products and services. Therefore, the EPDs for selected construction products that comply with this standard are only accounted for in the quantification. Secondly, although some EPDs provide data across the life-cycle stages, the site-specific conditions highly influence the impact values of materials from the transportation stage to the EoL. Hence, the waste data for the construction stage is collected from BRESmartWaste. This CW data collection platform gathers data from various buildings during their construction stage, estimates the material composition of waste, and categorises them based on their hazardous nature. Besides, this platform also presents the differences in the waste generation levels of each material based on building type (usage). So, these site-specific scenarios captured in the CW data collection platforms are incorporated into the ontology.

A spreadsheet is created to import the heterogeneous data collected from distributed data sources. Figure 5.24 shows the spreadsheet containing technical, generic, and LCA data of a few materials extracted from NORs. The collected data is converted into a machine-readable RDF format using the Cellfie plugin within the Protégé ontology editor. This plugin facilitates the conversion using the transformation rules. So, the transformation rules are developed to ease the conversion of technical and environmental impact data of selected materials into the ontology. Figure 5.25 displays the transformation rules developed to import instances of class *EPD_FireClayBricks*. The output of transformation rules is generated as axioms and added to the respective class of PRODCIRO (Figure 5.26). The same rule applies to other individuals created using Cellfie. In addition, a few individuals are created manually. This consistent and logical integration of classes, properties, and individuals in the PRODICRO will be used for further querying and integration with BIM platforms.

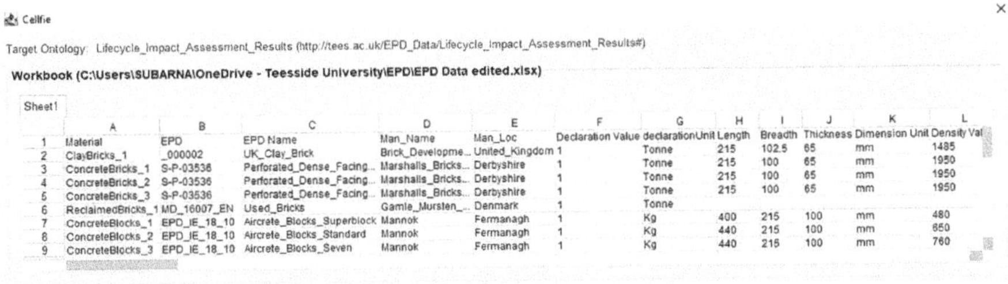

Figure 5.24 Data of Instances Imported Using Cellfie

Source: Drafted by the author

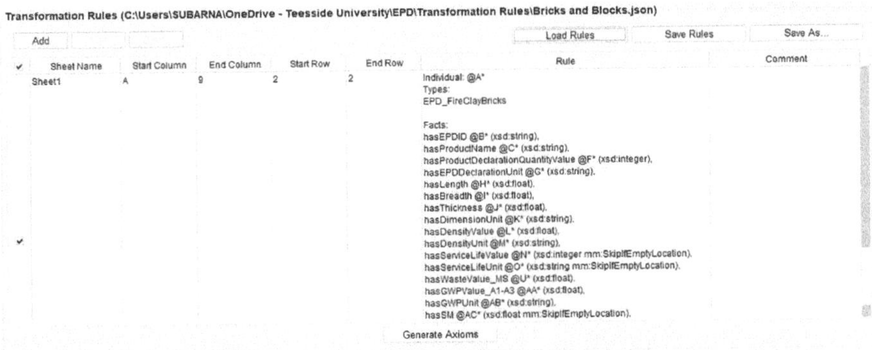

Figure 5.25 Transformation Rules to Import Instances of EPD_FireClayBricks Is Shown as an Example

Source: Drafted by the author

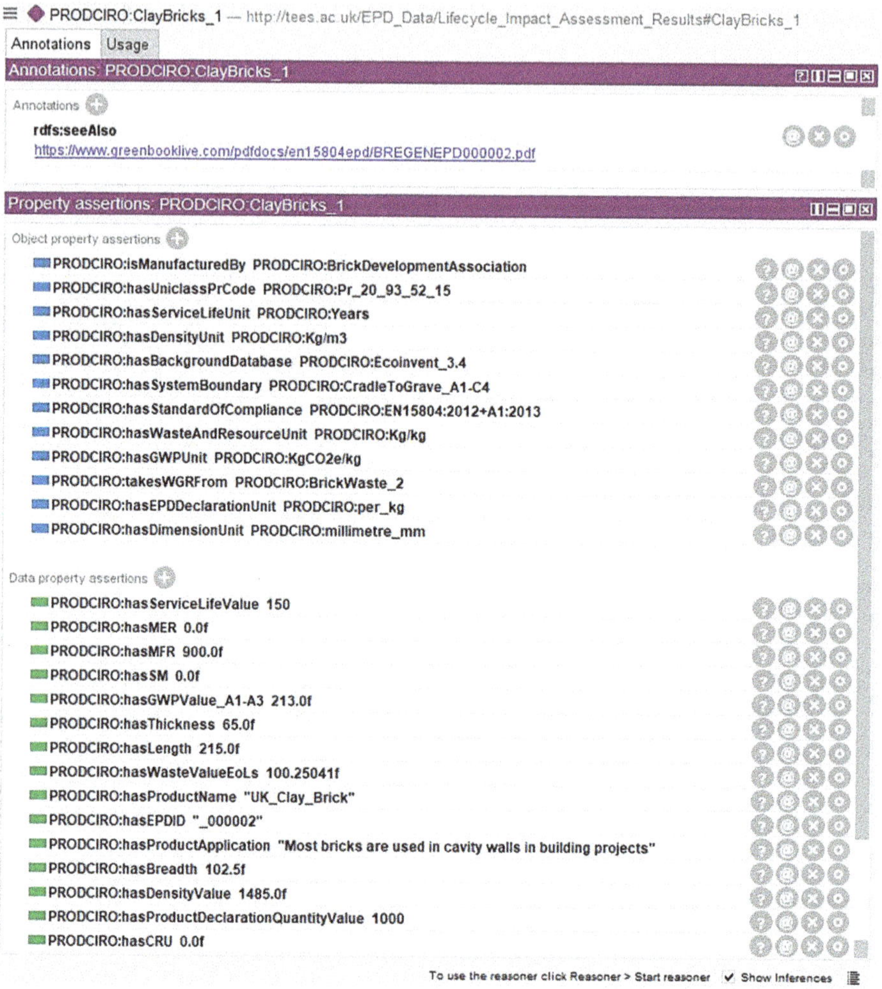

Figure 5.26 Imported Axioms Into PRODCIRO – ClayBricks_1 Is Shown as an Example

Source: Drafted by the author

5.3 PRODCIRO Evaluation

In addition to the technical phases, ontology evaluation is a supporting activity required to validate the consistency and accuracy of the defined classes and their relationships. The ontology evaluation techniques vary at each level of ontology development. In PRODCIRO, the syntactic evaluation of ontology is performed using a reasoner. The reasoner, semantic reasoner, or reasoning or rules engine helps infer implicit axioms from the asserted explicit facts [21]. Some examples of inferred axioms based on the ontology rules are presented in Sections 5.2.4 and 5.2.5. The overall evaluation of syntactic requirements of an ontology is evaluated using the Pellet reasoner that runs within the Protégé environment.

However, in addition to the existing evaluation technique, the application-based evaluation is considered appropriate to validate an ontology's vocabularies, concepts, data, hierarchies, and semantic relations [22]. It helps to decide the quality of the ontology based on the accuracy, consistency, and completeness of the results to the domain specification. Therefore, the application-based evaluation within a BIM environment using a test case is discussed in Chapter 6.

5.4 Knowledge Extraction From PRODCIRO

The next step followed by ontology development is to query the knowledge inferred from the ontology rules and check for its precisions. SPARQL is used in this research to query the developed ontology. The SPARQL query results are alternative ways to validate the ontology's precision. The web-based Fuseki-server API licensed by Apache is used in the research to implement and update SPARQL queries. The web-based Fuseki-server API licensed by Apache is used in the research to implement and update SPARQL queries. The below example (Figure 5.27) provides a sample query used to identify the product declaration quantity of individuals and unify it to weight-based units (mass) to ensure consistency in the CW results.

```
SELECT DISTINCT ?Products ?EPDID  ?PDQV ?PDQU ?Uniclass ?Length ?Breadth ?Thickness ?Density ?Mass ?ServiceLife
WHERE
{ ?Products a/rdfs:subClassOf* PRODCIRO:EPDProductType.
  BIND (STR (?Products) AS ?_ProductList)
  BIND (REPLACE (?_ProductList, "http://tees.ac.uk/EPD_Data/Lifecycle_Impact_Assessment_Results#", "", "i") AS ?ProductList)
  ?Products rdfs:label ?EPDID.
  ?Products PRODCIRO:hasProductDeclarationQuantityValue ?PDQV.
  ?Products PRODCIRO:hasEPDDeclarationUnit ?_DeclarationUnit.
  BIND (STR (?_DeclarationUnit) AS ?DeclarationUnit)
  BIND (REPLACE (?DeclarationUnit, "http://tees.ac.uk/EPD_Data/Lifecycle_Impact_Assessment_Results#", "", "i") AS ?PDQU)
  ?Products PRODCIRO:hasUniclassPrCode ?UniclassCode.
  BIND (STR( ?UniclassCode) AS ?_UniclassCode).
  BIND (REPLACE (?_UniclassCode, "http://tees.ac.uk/EPD_Data/Lifecycle_Impact_Assessment_Results#", "", "i") AS ?Uniclass)
 OPTIONAL {
  ?Products PRODCIRO:hasLength ?_ProductLength.
  ?Products PRODCIRO:hasBreadth ?_ProductBreadth.
  ?Products PRODCIRO:hasThickness ?_ProductThickness.
  ?Products PRODCIRO:hasDimensionUnit ?_DimensionUnit.}
  BIND (STR (?_DimensionUnit) AS ?DimensionUnit)
  BIND (REPLACE (?DimensionUnit, "http://tees.ac.uk/EPD_Data/Lifecycle_Impact_Assessment_Results#", "", "i") AS ?ProductDimensionUnit)
  BIND (COALESCE (IF (?ProductDimensionUnit = "metre_m", (?_ProductLength),
        IF (?ProductDimensionUnit = "centimetre_cm", (?_ProductLength / 100),
          IF (?ProductDimensionUnit = "millimetre_mm", (?_ProductLength/1000), "NA")))) AS ?Length)
  BIND (COALESCE (IF (?ProductDimensionUnit = "metre_m", (?_ProductBreadth),
        IF (?ProductDimensionUnit = "centimetre_cm", (?_ProductBreadth / 100),
          IF (?ProductDimensionUnit = "millimetre_mm", (?_ProductBreadth / 1000), "NA")))) AS ?Breadth)
  BIND (COALESCE (IF (?ProductDimensionUnit = "metre_m", (?_ProductThickness),
        IF (?ProductDimensionUnit = "centimetre_cm", (?_ProductThickness / 100),
          IF (?ProductDimensionUnit = "millimetre_mm", (?_ProductThickness / 1000), "NA")))) AS ?Thickness)
  OPTIONAL {
  ?Products PRODCIRO:hasDensityValue ?_ProductDensityValue.
  ?Products PRODCIRO:hasDensityUnit ?_DensityUnit.
  BIND (STR (?_DensityUnit) AS ?DensityUnit)
  BIND (REPLACE (?DensityUnit, "http://tees.ac.uk/EPD_Data/Lifecycle_Impact_Assessment_Results#", "", "i") AS ?_ProductDensityUnit)}
  BIND (COALESCE (?_ProductDensityValue, "NA") AS ?ProductDensityValue)
  BIND (COALESCE (?_ProductDensityUnit, "NA") AS ?ProductDensityUnit)
  OPTIONAL { ?Products PRODCIRO:hasProductMass ?_ProductMass. }
  BIND (COALESCE (IF (?ProductDensityUnit = "Kg/m3", ?_ProductDensityValue, (?ProductAreaDensityValue / ?ProductThickness)), "NA") AS ?Density)
  BIND (IF (?PDQU = "per_m3", (?PDQV * ?ProductDensityValue),
        IF (?PDQU = "per_m2", (?PDQV * ?ProductAreaDensityValue),
        IF (?PDQU = "per_unit", (?PDQV * ?_ProductMass),
        IF (?PDQU = "per_tonne", (?PDQV * 1000),
        IF (?PDQU = "per_kg", (?PDQV), "NA"))))) AS ?Mass)
 OPTIONAL {
  ?Products PRODCIRO:hasServiceLifeValue ?_ServiceLife.
  ?Products PRODCIRO:hasServiceLifeUnit ?ServiceLifeUnit.}
  BIND (COALESCE (?_ServiceLife, "NA") AS ?ServiceLife)
```

Figure 5.27 Sample SPARQL for Knowledge Extraction From the Ontology

Source: Drafted by the author

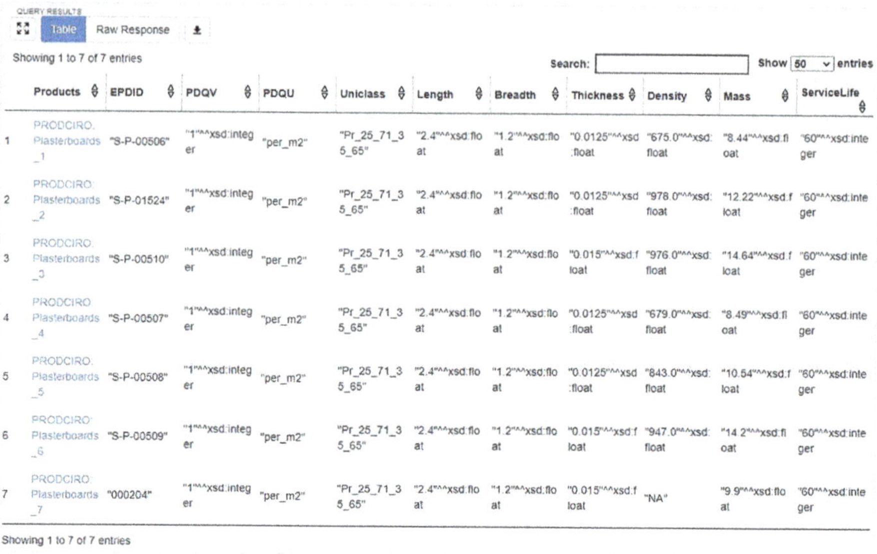
QUERY RESULTS

Table | Raw Response

Showing 1 to 7 of 7 entries

Search: Show 50 entries

	Products	EPDID	PDQV	PDQU	Uniclass	Length	Breadth	Thickness	Density	Mass	ServiceLife
1	PRODCIRO:Plasterboards_1	"S-P-00506"	"1"^^xsd:integer	"per_m2"	"Pr_25_71_35_65"	"2.4"^^xsd:float	"1.2"^^xsd:float	"0.0125"^^xsd:float	"675.0"^^xsd:float	"8.44"^^xsd:float	"60"^^xsd:integer
2	PRODCIRO:Plasterboards_2	"S-P-01524"	"1"^^xsd:integer	"per_m2"	"Pr_25_71_35_65"	"2.4"^^xsd:float	"1.2"^^xsd:float	"0.0125"^^xsd:float	"978.0"^^xsd:float	"12.22"^^xsd:float	"60"^^xsd:integer
3	PRODCIRO:Plasterboards_3	"S-P-00510"	"1"^^xsd:integer	"per_m2"	"Pr_25_71_35_65"	"2.4"^^xsd:float	"1.2"^^xsd:float	"0.015"^^xsd:float	"976.0"^^xsd:float	"14.64"^^xsd:float	"60"^^xsd:integer
4	PRODCIRO:Plasterboards_4	"S-P-00507"	"1"^^xsd:integer	"per_m2"	"Pr_25_71_35_65"	"2.4"^^xsd:float	"1.2"^^xsd:float	"0.0125"^^xsd:float	"679.0"^^xsd:float	"8.49"^^xsd:float	"60"^^xsd:integer
5	PRODCIRO:Plasterboards_5	"S-P-00508"	"1"^^xsd:integer	"per_m2"	"Pr_25_71_35_65"	"2.4"^^xsd:float	"1.2"^^xsd:float	"0.0125"^^xsd:float	"843.0"^^xsd:float	"10.54"^^xsd:float	"60"^^xsd:integer
6	PRODCIRO:Plasterboards_6	"S-P-00509"	"1"^^xsd:integer	"per_m2"	"Pr_25_71_35_65"	"2.4"^^xsd:float	"1.2"^^xsd:float	"0.015"^^xsd:float	"947.0"^^xsd:float	"14.2"^^xsd:float	"60"^^xsd:integer
7	PRODCIRO:Plasterboards_7	"000204"	"1"^^xsd:integer	"per_m2"	"Pr_25_71_35_65"	"2.4"^^xsd:float	"1.2"^^xsd:float	"0.015"^^xsd:float	"NA"	"9.9"^^xsd:float	"60"^^xsd:integer

Showing 1 to 7 of 7 entries

Figure 5.28 Sample Output of SPARQL

Source: Drafted by the author

The SPARQL has four query forms/clauses: SELECT, CONSTRUCT, ASK, and DESCRIBE. The use of these query forms depends on their application. In this sample, the SELECT query form matches patterns in the ontology to return results. The variables within the SELECT clause represent the list of information to be displayed after executing the query.

The inferred results of the query are presented in Figure 5.28. The results give the list of EPDs available for plasterboard and the associated generic and technical details. The use of SPARQL required for quantifying CW and ranking alternatives is discussed in Chapter 6, and the workflow is demonstrated by using a test case building.

5.5 Summary

Aligning with the DSR approach, the application of the proposed CWQ framework is demonstrated using an ontology. The ontology development comprises specification, conceptualisation, formalisation, and implementation. They are further categorised into seven steps by SKEM, which is adopted in the research. The highlights of the chapter include the following.

- The main aim of PRODCIRO development is to create an interoperable CW data model with standardised and consistent semantics. The outputs of PRODCIRO support design team members in making informed CW decisions from the early design stages.

- The primary data sources that supported the research are BRESmartWaste reports, product-specific EPDs from the manufacturer, Uniclass, and EWC. The guidelines from CLC and RICS are also the primary knowledge sources.
- The concepts defined in BMO, MAT, and QUDT are integrated into PRODCIRO.
- In addition to the selected knowledge sources, the important entities of PRODCIRO are extracted from NORs. The NORs are the outputs of the literature review, semi-structured interviews and exploratory data analysis.
- There are eight main classes within PRODCIRO. It includes *EPDProductType, EPDProductData, BuildingWasteMaterialType, BuildingType, BuildingElements, EuropeanWasteCatalogue_EWC, UniclassPrCodes,* and *Unit.*
- The object properties define the relationship between the instances of two classes, and the data properties provide the relationship between an instance and an attribute data. The relationships are structured as triples: subject, predicate, and object.
- The instances are the individual entities within a specific class type. The individuals within PRODCIRO hold generic, technical, waste, and carbon data for materials. The individuals of each class decide the CW generation and its impact levels to facilitate optimisation.
- The Pellet reasoner supported the syntactic and semantic validation of the ontology.
- SPARQL queries are used to extract knowledge from PRODCIRO.
- The design and development of BIM-integrated UI and validation of the framework are discussed in Chapter 6.

References

1. P. Pauwels, S. Zhang and Y.-C. Lee, Semantic web technologies in AEC industry: A literature overview. *Automation in Construction*, 2017. 73: pp. 145–165. https://doi.org/10.1016/j.autcon.2016.10.003
2. N.F. Noy and D.L. McGuinness, *Ontology development 101: A guide to creating your first ontology*, 2001. Stanford Knowledge Systems Laboratory. https://protege.stanford.edu/publications/ontology_development/ontology101.pdf. Last access: 08 December 2024.
3. NBS, *Uniclass – Unified construction classification*, 2015. https://www.thenbs.com/our-tools/uniclass. Last access: 08 December 2024.
4. CLC, *Zero avoidable waste in construction*, 2020. https://www.constructionleadershipcouncil.co.uk/wp-content/uploads/2016/05/ZAW-Report-Final-Draft-25-February–2020.pdf. Last access: 15 May 2023.
5. RICS, *Whole life carbon assessment for the built environment*, 2023. https://www.rics.org/profession-standards/rics-standards-and-guidance/sector-standards/construction-standards/whole-life-carbon-assessment. Last access: 08 December 2024.
6. Environment Agency, *Technical guidance WM3: Waste classification – Guidance on the classification and assessment of waste*, 2014. https://www.gov.uk/government/publications/waste-classification-technical-guidance. Last access: 08 December 2024.
7. M. Gruninger, Methodology for the design and evaluation of ontologies, in *Proc. IJCAI'95, Workshop on basic ontological issues in knowledge sharing*, 1995. https://www.researchgate.net/publication/2288533_Methodology_for_the_Design_and_Evaluation_of_Ontologies. Last access: 15 December 2024.
8. E. Simperl, Reusing ontologies on the Semantic Web: A feasibility study. *Data & Knowledge Engineering*, 2009. 68(10): pp. 905–925. https://doi.org/10.1016/j.datak.2009.02.002

9. P. Pauwels and W. Terkaj, EXPRESS to OWL for construction industry: Towards a recommendable and usable ifcOWL ontology. *Automation in Construction*, 2016. 63: pp. 100–133. https://doi.org/10.1016/j.autcon.2015.12.003
10. M.H. Rasmussen, M. Lefrançois, G.F. Schneider and P. Pauwels, BOT: The building topology ontology of the W3C linked building data group, in *Semantic web*, 2021. pp. 143–161. https://w3c-lbd-cg.github.io/bot/. Last access: 06 December 2024.
11. P. Pauwels and A. Roxin, SimpleBIM: From full ifcOWL graphs to simplified building graphs, in *eWork and eBusiness in architecture, engineering and construction*, 2017. CRC Press. pp. 11–18. ISBN: 1315386909.
12. P. Valluru, J. Karlapudi, K. Menzel, T. Mätäsniemi and J. Shemeikka, A semantic data model to represent building material data in AEC collaborative workflows, in *Boosting collaborative networks 4.0: 21st IFIP WG 5.5 working conference on virtual enterprises, PRO-VE 2020, Valencia, Spain, November 23–25, 2020, Proceedings 21*, 2020. Springer. https://doi.org/10.1007/978-3-030-62412-5_11
13. S.C.-F. María Poveda-Villalón, *Material properties ontology*, 2021. Ontology Engineering Group. https://bimerr.iot.linkeddata.es/def/material-properties/. Last access: 10 March 2023.
14. M. Fernández-López, A. Gómez-Pérez and N. Juristo, *Methontology: From ontological art towards ontological engineering*, 1997. https://cdn.aaai.org/Symposia/Spring/1997/SS-97-06/SS97-06-005.pdf. Last access: 10 September 2023.
15. B.C. Villazón-Terrazas, M. Suárez-Figueroa and A. Gómez-Pérez, A pattern-based method for re-engineering non-ontological resources into ontologies. International *Journal on Semantic Web and Information Systems (IJSWIS)*, 2010. 6(4): pp. 27–63. https://doi.org/10.4018/jswis.2010100102
16. L. Prévot, C.-R. Huang, N. Calzolari, A. Gangemi, A. Lenci and A. Oltramari, *Ontologies and the Lexicon: A multi-disciplinary perspective*, 2010. Cambridge University Press. ISBN: 9780521886598.
17. D. Allemang and J. Hendler, *Semantic web for the working ontologist: Effective modeling in RDFS and OWL*, 2011. Elsevier. ISBN: 9780123859662.
18. M. Uschold and M. Gruninger, Ontologies: Principles, methods and applications. *The Knowledge Engineering Review*, 1996. 11(2): pp. 93–136. https://doi.org/10.1017/S0269888900007797
19. S. Sivashanmugam, S. Rodriguez, F.P. Rahimian, F. Elghaish and N. Dawood, Enhancing information standards for automated construction waste quantification and classification. *Automation in Construction*, 2023. 152: p. 104898. https://doi.org/10.1016/j.autcon.2023.104898
20. W3C, *OWL 2 web ontology language – Document overview* (Second Edition), 2009. https://www.w3.org/TR/owl2-overview/. Last access: 20 March 2023.
21. S. Abburu, A survey on ontology reasoners and comparison. *International Journal of Computer Applications*, 2012. 57(17): pp. 33–39. https://doi.org/10.5120/9208–3748
22. J. Brank, M. Grobelnik and D. Mladenic, *A survey of ontology evaluation techniques*, 2005. Citeseer. pp. 166–170. https://aile3.ijs.si/dunja/SiKDD2005/Papers/BrankEvaluationSiKDD2005.pdf. Last access: 06 December 2024.

6 Building Waste Tool – Design, Development, and Validation

6.1 Introduction

The PRODCIRO is integrated with a BIM platform to automatically extract design and material parameters from the BIM model, reduce the manual work associated with developing long SPARQL queries, and enable seamless communication with BIM authoring tools. A BIM-based plugin is developed to demonstrate the integration. The plugin is named 'Building Waste Tool' (BWT). The workflow adopted in the BWT ensures that the process of quantifying waste and informing optimal material choices is automated by enabling seamless and dynamic information flows between BIM and the PRODCIRO. The Dynamo Revit within the Autodesk Revit environment and RDF Library (RDFLib) supported the development of BWT. The plugin consists of seven components grouped as three panels: (i) 'Home page' that consists of three components including 'About Building Waste Tool', 'Using Building Waste Tool', and 'Prerequisites', (ii) 'Building Waste Assessment' consisting of three components, including 'Create Custom Waste Parameters', 'Perform Waste Analysis', and 'View Waste Reports', and (iii) 'Additional Information' with 'Waste and Carbon Data' component. The buttons are created in the Autodesk Revit environment using Dyno Browser to enhance user interaction with these components. The overall workflow is validated using a real-life test case building. The quantification results are compared with other CW datasets to validate the accuracy of the results, compare the differences in CW values, and identify the data gaps.

6.2 Software Modelling Environment for BWT

This section overviews BIM software environments and RDF libraries that supported the BWT development.

6.2.1 Dynamo Revit and RDFLib

The value of innovative frameworks is decided by their integration and interoperability levels with BIM platforms. The literature also claims that CW platforms with seamless communication with the BIM environment would support waste-driven

DOI: 10.1201/9781003610762-6

decision-making across the life cycle stages [1, 2]. Also, this integration further improves the standardisation and consistency of the CW data and overcomes the information gaps limiting the application of the CW platforms [3–5]. So, the PRODCIRO is integrated with the BIM platform to access the standardised data required for quantifying CW and informing optimal material choices. Besides, one of the downsides of ontology applications is the need to write long SPARQL queries to extract relevant knowledge. So, these two factors emphasise the importance of RDF and BIM integration through a correspondence medium that facilitates seamless information flows and enhanced communication.

Based on the literature review, it was identified that the interoperability of BIM with external platforms would enable this integration using APIs and add-ins (discussed in Section 3.3.4). The APIs facilitated by the BIM authoring platforms mostly complain with .NET languages like C# and VB.NET. In addition, the Autodesk Revit BIM platform (the selected tool for this research) offers an add-in called Dynamo Revit that runs with Python language. Among two API options, the BWT used Dynamo Revit 2.10 to integrate the semantic CW data model with Autodesk Revit 2022. Besides, the RDFLib also runs with a Python library, which makes it compatible with the Dynamo Revit environment. The library also contains parsers and serialisers for RDF/XML, Turtle, N3, N-Triples, and JSON-LD. It also provides SPARQL API extensions to enhance the querying mechanism of ontologies, which is the primary functionality utilised in this research by integrating PRODCIRO with BIM. Therefore, the functionalities and compatibility of RDFLib with the Dynamo Revit environment facilitate the BWT to automatically extract, process, and optimise BIM design and material information based on the CW impact results.

6.2.2 UI Development Software

Further to integration, the functionality of BWT is enhanced by developing an interactive UI within the Autodesk Revit environment. The interface is facilitated by Dyno, a browser that helps to organise and deploy dynamo codes into the Autodesk Revit workspace. The tool allows the developer to convert the dynamo scripts as ribbons. Therefore, the functionalities of the Dyno browser are incorporated into the research to develop an interactive UI via a plugin. This interface, therefore, significantly improves the speed, automation, engagement, collaboration, and communication among the project team. Besides, it facilitates users in comparing multiple design options, benchmarking WGRs, and making informed decisions with a single click.

6.3 BWT Design and Development

The primary functions expected from the proposed framework are to enhance standardisation, automation, and interoperability between CW and BIM platforms. These three pillars promise a collaborative data environment that helps the design team members integrate CW factors into decision-making. In addition to the benefits the ontology provides, a UI becomes essential to do the following:

i) Allow users to select specific or multiple elements within an element family to conduct the analysis.
ii) Extract design and material information for the selected element(s) from the BIM model.
iii) Explore the impacts of each material before making choices.
iv) Identify the type of alternatives available to replace a specific material.
v) Understand the manufacturers, their manufacturing units, the type of materials they manufacture and the transportation distance to the site.
vi) Map the material information with CW information without human intervention.
vii) Compare multiple design options with reduced redundancies of BIM models.

This section discusses the seven components of BWT and their associated process. Figure 6.1 depicts an overview of the BWT workflow integrating the proposed components. The Dynamo Revit integrated with RDFLib does the following: extract design and material information from the BIM model, assign Uniclass codes to elements and materials within BIM, filter and map BIM and ontology data based on the Pr codes, quantify CW, optimising materials based on CW impact results, present the results to the users, and embed the CW information into the BIM model. The dynamo graphs developed to do the above functions are grouped and converted as Autodesk Revit buttons to enhance the interaction of the imported ontology.

Figure 6.2 presents the components that are represented as buttons in the Autodesk Revit interface. The components within the 'Home Page' and 'Additional

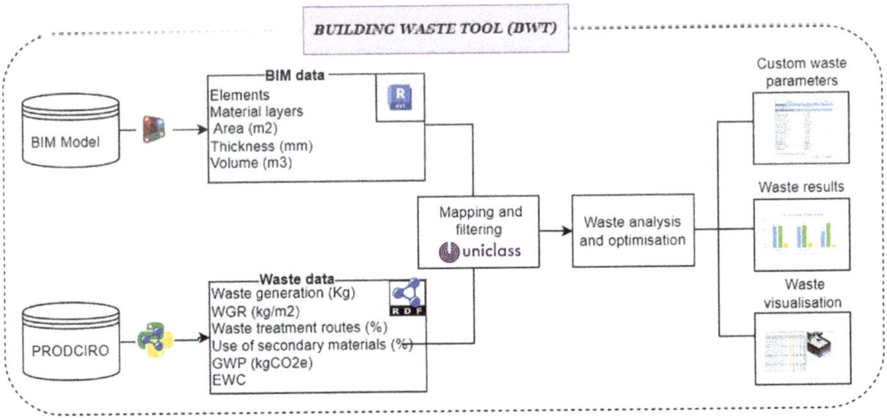

Figure 6.1 BWT Workflow

Source: Drafted by the author

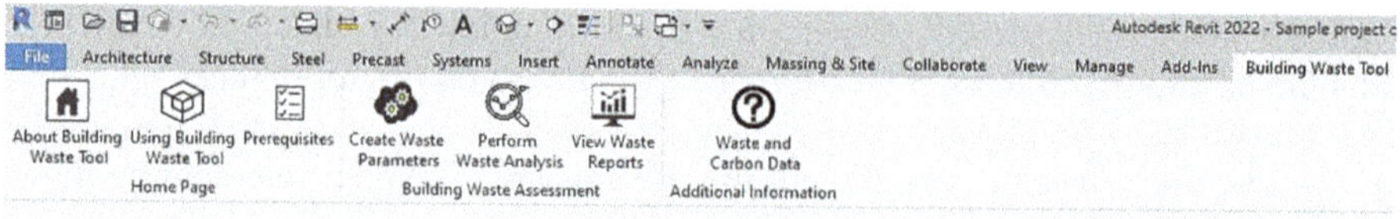

Figure 6.2 BWT Ribbon

Source: Drafted by the author

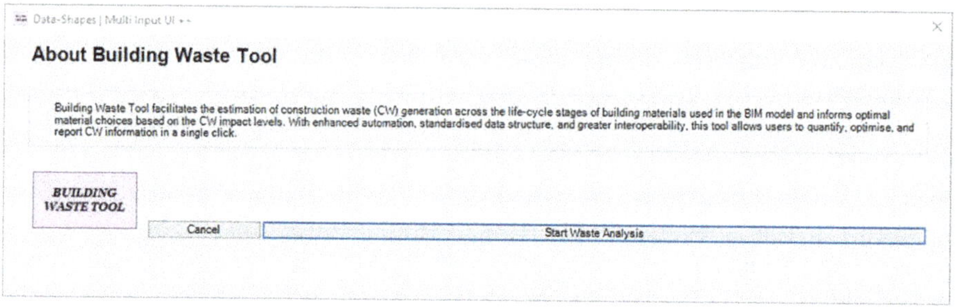

Figure 6.3 About BWT

Source: Drafted by the author

Information' panels are independent and have no dependencies. However, the second and third components in the 'Building Waste Assessment' panel have some dependencies over the previous ones. This implies that the second component is dependent on the first, and the third is dependent on the first and second components. If the user wants to 'Perform Waste Analysis' (second component), the initial step is to 'Create Waste Parameters' (first component) to import the waste data from the analysis. Moreover, if the user wants to 'View Waste Reports' (third component), the primary steps are to 'Create Waste Parameters' (first component) and 'Perform Waste Analysis' (second component).

6.3.1 BWT Panel 1 – 'Home Page'

The 'Home Page' panel of the BWT consists of three components: (a) 'About Building Waste Tool', (b) 'Using Building Waste Tool', and (c) 'Prerequisites'. The functions of each component are as follows:

a) **About Building Waste Tool**: This component describes the features and benefits of the BWT. Figure 6.3 displays the information provided in the home page button.
b) **Using Building Waste Tool**: This component provides information about the steps associated with selecting building elements, quantifying waste, and obtaining optimal materials based on the quantified CW value via BWT (Figure 6.4).

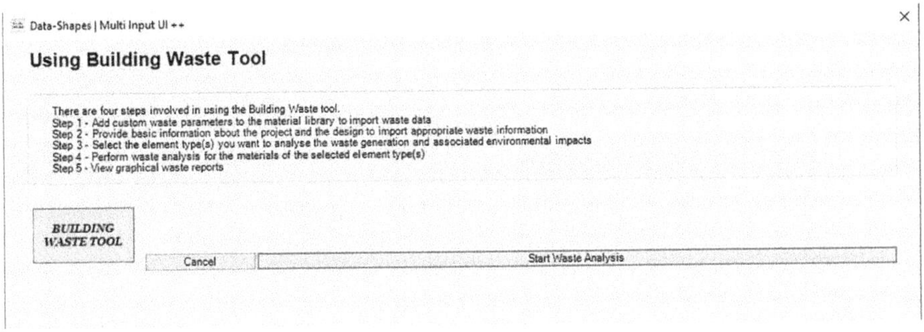

Figure 6.4 Using BWT

Source: Drafted by the author

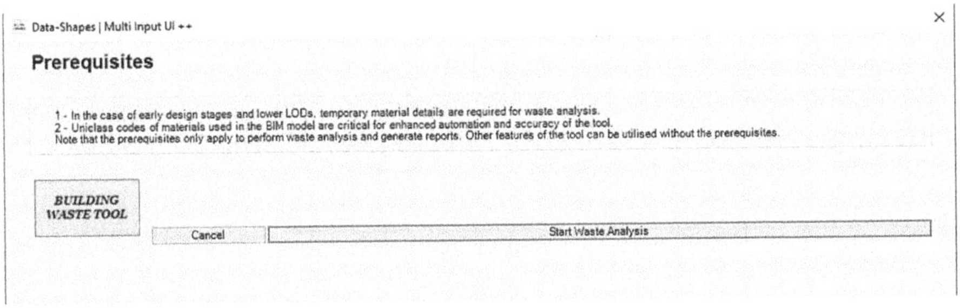

Figure 6.5 Prerequisites for Using BWT

Source: Drafted by the author

a) **Prerequisites**: This component provides the prerequisites vital for extracting appropriate information and displaying accurate CW quantification results using BWT (Figure 6.5).

6.3.2 *BWT Panel 2 – 'Building Waste Assessment – Create Waste Parameters'*

The parameters in the BIM model allow users to control information and properties of various model elements and categories. In addition to default parameters, custom parameters can be created to support building performance assessments and simulations. The two general types of parameters are type and instance parameters. The significant difference is that the change in the type parameter affects the whole family, while the instance parameter influences only the selected objects or instances. These parameters may be specific to a project, specific to a family type or shared across multiple projects. In this case, the custom waste parameters are applied to the material category within the project, which acts as an 'instance project parameter of a material(s)' in the BIM environment. Furthermore, these parameters are grouped under 'Green Building Properties'.

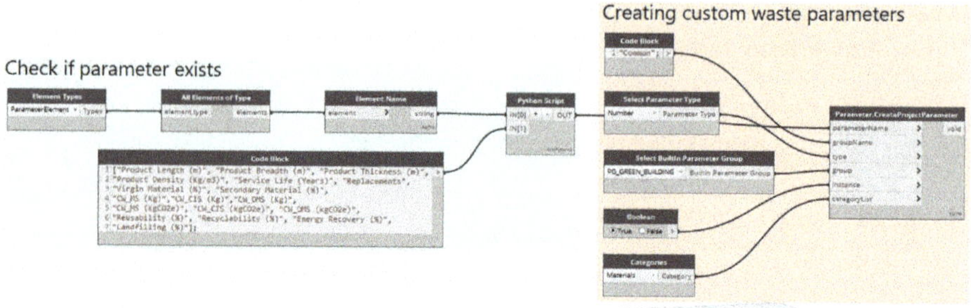

Figure 6.6 Adding Custom Waste Parameters

Source: Drafted by the author

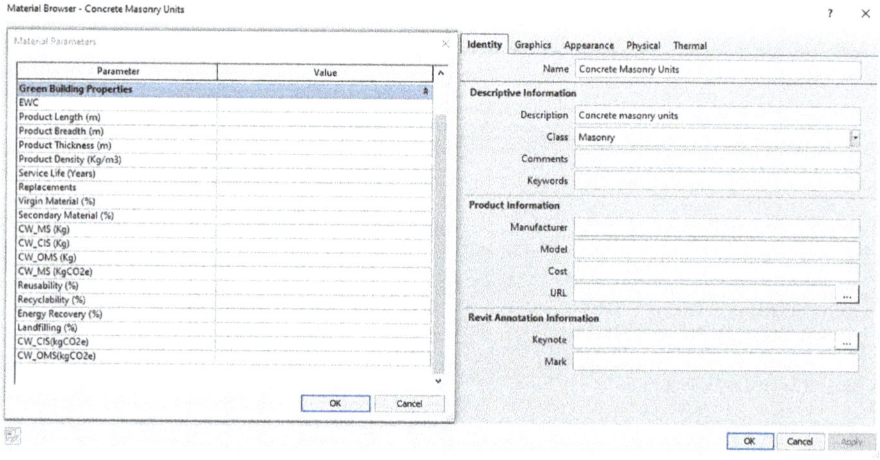

Figure 6.7 Sample Custom Waste Parameters Added to the BIM Model

Source: Drafted by the author

The 'Create Waste Parameters' component allows the creation of multiple custom parameters automatically using Dynamo Revit. Figure 6.6 shows the process of adding a sample waste parameter to the material category. The custom parameters rely on the information accounted for as CW indicators in the literature and interviews, which are then conceptualised and formalised in the PRODCIRO (Section 5.2). Figure 6.7 presents the list of custom CW parameters added to the Concrete Masonry Units (representing the CW indicators obtained from exploratory data analysis in Section 4.5.2).

6.3.3 *BWT Panel 2 – 'Building Waste Assessment – Perform Waste Analysis'*

This component allows users to quantify the CW generation levels and optimise building materials that generate less waste and carbon throughout their life-cycle stages. There are eight steps associated with the CWQ workflow. The functions of each step will be discussed individually in this section. However, all these steps are integrated into a single visual graph to create seamless data transfer between BIM and PRODCIRO. The aggregated graph acts as a 'Perform Waste Analysis' component in BWT. The steps include the following.

6.3.3.1 Step 1 – Obtain Basic Information About the Project and the Design

The primary step in the BWT is to obtain basic information about the selected BIM design (Figure 6.8). Among the four inputs, the indicators 'Building GFA (m^2)' and 'Building Usage Type' are critical to extracting appropriate information from the semantic CW model. Since the WGR of material during the construction phases are scenario-based and varies with building usage, these two indicators are incorporated into the analysis (explained in Step 5).

While the other two indicators, 'Project Name' and 'Design Name', have no direct influence on data, they are used to compare different design options and generate CW reports. These pieces of information are prerequisites to facilitate the CWQ process.

6.3.3.2 Step 2 – Select Element(s) of a Specific Family Type From the BIM Design

BIM's parametric modelling and visualisation features (discussed in Section 3.3.3) play a vital role in this research, reducing the time and effort required to incorporate material changes in the BIM models. Applying these functionalities, a UI is created to allow users to visualise the whole building and select building element(s) that must be analysed using the BWT (Figure 6.9). Changes in a selected building element are reflected across other elements of the same family type using parametric modelling functions, thus reducing human intervention.

The interface allows the user to select single or multiple elements from the same family of their choice to quantify waste arising from each material layer. It

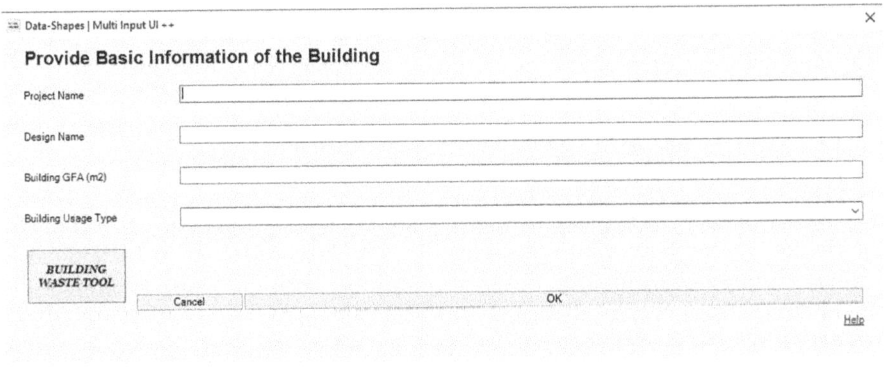

Figure 6.8 UI – Basic Information of the Building

Source: Drafted by the author

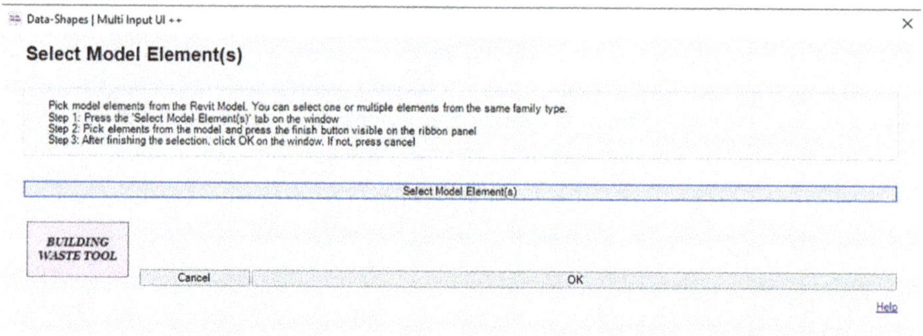

Figure 6.9 UI – Select Model Element(s)

Source: Drafted by the author

is important to note that the prerequisite before undertaking this step is to assign the Uniclass code(s) for the element and its material layers. The rationale for using Uniclass and its relationship with other entities in the PRODCIRO are explained in Section 5.2.5.

6.3.3.3 Step 3 – Extract the Material Information of the Element Layers

The information in PRODCIRO is specific to materials, not the element itself. This means that the CW generation levels and other influencing indicators are assigned at the material level to ensure granularity in the results. So, the materials associated with the element, including its name and dimensional properties like area, volume, and thickness, and Uniclass codes are extracted to quantify the waste generation levels and inform optimal material choices. Besides, a building element could have either single or multiple material layers. So, the position of each material layer within the element is extracted to assess and replace the existing material with an optimised one at the same position. The entire process is underpinned by BIM's life-cycle information management and automated quantity estimation attributes (discussed in Section 3.3.1). The extracted BIM data facilitates the user to filter, map, and optimise alternatives for existing materials in the BIM model.

6.3.3.4 Step 4 – Filter and Map Building Products Based on Their Application and Uniclass Codes

A filtering process is applied to select appropriate materials for each element. Firstly, the alternatives within the semantic CW data model could be applied to various building elements. So, the relationship between the material alternatives and building element(s) is established using the ontology's *applicableForBuilding-Elements* object property (discussed in Section 5.2.5). The Uniclass EF codes are used to classify the building elements. Hence, using this relationship, the SPARQL query filters the materials suitable for the specific building element.

Table 6.1 Uniclass Code Structure

Uniclass code	*Description*	*Hierarchy*
Pr_20	Structure and general products	Group
Pr_20_93	Unit structure and general products	Sub-group
Pr_20_93_52	Masonry walling units	Section (Used for material mapping)
Pr_20_93_52_15	Clay Bricks	Object (used for assigning the codes to the PRODICRO classes)

Source: Drafted by the author

Secondly, the alternative choices from the *EPDProductType* class are mapped with the BIM material data using the Uniclass Pr codes. The workflow utilises the Pr codes assigned to the materials in the BIM (a prerequisite in BWT as discussed in Section 6.3.1) and PRODCIRO (under the has*UniclassPrCode* property as discussed in Section 5.2.5) to filter alternatives that apply to the specific material. The Pr codes follow a four-tier hierarchical approach (Table 6.1).

From the four tiers, the section-based classification (third tier) is adopted in the optimisation process to broaden the material search within the model. For instance, if clay brick (Pr_20_93_52_15) is used as a material in BIM, the user will be prompted with alternatives that are not just specific to clay bricks but also the ones that come under the masonry walling unit's section (Pr_20_93_52). These filters allow the tool to pick materials that are technically suitable for the selected element type before conducting the quantification. These factors are integrated because ranking and optimising materials by comparing their environmental impacts becomes reliable only if the associated technical parameters are consistent.

6.3.3.5 *Step 5 – Quantify, Rank, and Optimise Materials Based on Their Waste Impact Levels*

Unify the declared units:

Based on the collected data, it was clear that the declared units of waste generation and life-cycle impact factors across CW reports and EPDs are non-standardised and inconsistent. For instance, the EPDs declare life-cycle impact factors per material area, volume, mass, or quantity, limiting its direct application in CWQ. Utilising these inconsistent units will cause higher uncertainties in the quantification results and comprise optimisation accuracy. Therefore, the waste generation and life-cycle impact factors are unified to 'per 1 kg of the product', and the results are aggregated under the 'ProductMass' property to ensure accuracy, automation, and consistency across the complete workflow.

Quantify waste generation across the life-cycle stages:

To ensure granularity in the waste results, the life-cycle stages are classified as product manufacturing, construction/installation, operation and maintenance, and EoL stages, as emphasised by the LCA framework (discussed in Section 3.6.2). No single source of data and methodology is currently available to estimate CW generation across all these stages. So, specific methods that suit the data needs are adopted. Firstly, the waste generated during the manufacturing stage is product-based and comes directly from the manufacturer's source, so the data is directly extracted from the EPDs after unifying the declaration units. Unlike the manufacturing stages, the waste generation during construction/installation, operation and maintenance, and EoL stages is scenario-based. This means that no products of the same type can generate the same amount of waste during construction/installation, operation and maintenance, demolition/deconstruction, or waste processing because of the diverse waste factors that influence their generation rates during those stages. Therefore, each stage is analysed separately, and an appropriate method supporting the CWQ is selected.

The CW reports from BRESmartWaste are used to measure the waste generation during the construction/installation stages based on the given design. The waste generated during this stage is influenced by several variables like building usage type, structural frame, construction methods, procurement, and other external factors (discussed in Section 2.7.2). However, the aggregated reports from BRESmartWaste show that the building usage type and geographical location are the most influencing variables in the construction/installation stages. So, the waste generation data and its relationship with influencing variables are gathered from BRESmartWaste reports to estimate the expected waste levels during the construction and installation of the building.

Despite the operation and maintenance stage of the building holding the longest span, no databases provide numerical data on waste generation and its influencing variables during this stage. So, the lifetime analysis approach is adopted. The reference service period (RSP) of the building and the designed life of building products are used to quantify the expected level of waste generation during the operation and maintenance stages. The RICS LCA guidance suggests that the RSPs for domestic and non-domestic buildings in the UK are assumed to be 60 years [6] while conducting LCA. Besides, the RSP of individual building components incorporated into PRODCIRO is also used to filter materials. Although this value might differ between buildings due to various influential factors like technical, environmental, economic, and political factors, the base value is used to benchmark the CW performance.

The assumed RSP and the expected service life of each building product given in the EPDs will be used to estimate the number of replacements required for a product across the 60 years. The replacements decide the quantity of CW expected to be generated during the operation and maintenance stage of the building. For instance, if the service life of the product given in the EPDs is 20 years, it is assumed that the material will undergo two replacements during the building operation and maintenance stages. The third replacement will be when the building reaches its end of RSP, so it is categorised as EoL impacts. The EoL stages are currently not included in the research.

Figure 6.10 Screenshot of SPARQL to Assess CW During the Operation and Maintenance Stages

Source: Drafted by the author

Table 6.2 LCA Phases Reflected in CWQ

Goal and scope	• Goal – Quantify CW generation of developed design and inform optimal material choices based on CW and its impact levels. GWP is the indicator used to measure the impact of CW. • System Boundary – Product manufacturing (material extraction, manufacturing, and transportation), construction/installation, and in-use (operation and maintenance works) stages. • Functional Unit – 1 kg of estimated product quantity
Inventory	• Input (design and material details) data – BIM model • Output data – BRESmartWaste CW reports and EPDs • Other classification and guidance documents (as highlighted in Section 5.2.1).
Impact assessment (CWQ)	CW is categorised according to each building element's material type and life cycle stages.
Interpretation and decision-making	The quantified CW data is expected to allow users to do the following: • Identify the hotspots like the type of material generating a high volume of waste • Determine the relationship between the CW generation levels and the associated GWP impacts • Validate the estimated values with real-life scenarios • Propose opportunities (design changes and WM strategies) to minimise the CW generation levels.

Source: Drafted by the author

In specific scenarios where the assumed service life of an individual product is higher than RSP, the material is expected to be removed at the end of the RSP of the building. Thus, in such scenarios, the product's service life is assumed to be equal to RSP. Other life-cycle stages after the operation and maintenance stages are excluded due to a lack of site-specific CW data (Figure 6.10). Table 6.2 shows the life cycle stages covered in the research, along with its data sources and interpretation methods.

6.3.3.6 *Assess and Rank Materials Based on the GWP values*

The quantified CW generation levels and their associated impacts decide the rank of materials. As discussed in the literature (Sections 2.7.4 and 4.4.5.3), impact-based analysis is essential to reduce the environmental impacts of CW. Thus, GWP values of waste materials are assessed to report the waste impacts. A material with low

CW impact (measured as GWP) is placed at the top, followed by other materials. Despite the flexibility of the tool across the life-cycle stages, only the embodied carbon of materials associated with product manufacturing phases is included in the research due to data limitations within the life-cycle impact data inventories. Therefore, only the aforementioned CW impact factors influence the ranking and optimisation of materials.

6.3.3.7 Step 6 – Create New Material(s) With CW Information and Import Them Into the BIM Material Library

Based on the ranking results, the top-ranked material will be imported into the BIM library as a new material. The information acquired from the semantic CW data model, including its technical, waste, and impact information, is also imported as custom parameters. This process is incorporated into the BWT to allow designers to select similar material(s) for other elements without repeating Steps 1 to 5, utilising the parametric modelling functionalities of BIM. Besides, integrating CW data into BIM models would enhance the efficiency of waste analysis in the design stages.

6.3.3.8 Step 7 – Create a Duplicate Family for Selected Elements

In order to minimise data errors and allow users to compare the waste difference between the baseline and optimised design, the user-selected element is given a new family name. This new family duplicates the existing one along with the material layers. Besides, it is also a preferred way of dealing with families should the parameter changes be significant.

6.3.3.9 Step 8 – Replace Existing Materials of the Duplicate Family With Newly Created Materials

The newly created material(s) and associated parameters replace the existing material(s) of the created duplicated family. If multiple elements of the same family are selected, the original family name is changed to the duplicate family name, and Steps 5 and 6 will be automatically implemented. The waste results of the optimised materials are presented as a CW schedule within the project library. Besides, the estimated waste results stored as custom parameters could be used for design optimisation, supply-chain decision-making, procurement, and preparing waste resources on-site.

6.3.4 BWT Panel 2 – 'Building Waste Assessment – View Waste Reports'

This component allows the visualisation of waste results as graphs. Since there are multiple waste parameters, presenting them in a single graph is challenging for the users. So, the interface allows the user to select a specific parameter to visualise and present the results in a graphical format. Figure 6.11 shows the UI that allows the user to select a parameter to be selected for generating the graphs.

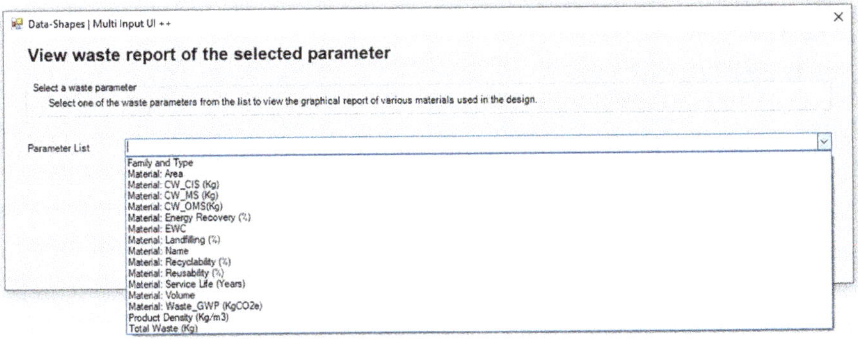

Figure 6.11 UI – View Waste Reports

Source: Drafted by the author

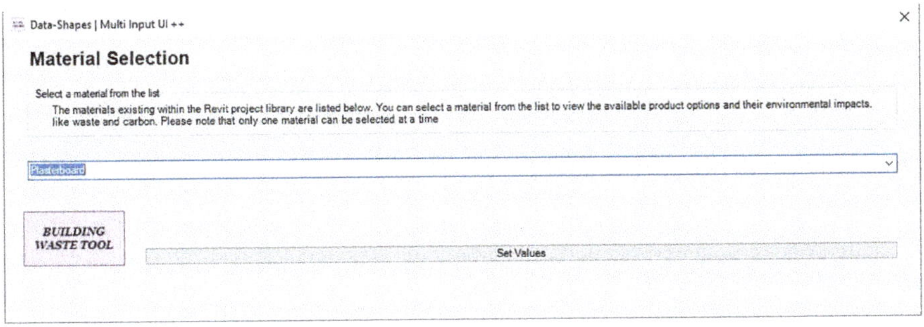

Figure 6.12 UI – Material Selection

Source: Drafted by the author

6.3.5 *BWT Panel 3 – 'Additional Information – Waste and Impact Data'*

The final component provides the complete list of building products added to PRODCIRO with details on their technical, waste, and impact values. The user is prompted with an interface to select one of the materials existing within the Autodesk Revit material library. Figure 6.12 shows the UI with plasterboard as a selected material. Based on the user inputs, the list of available alternative material choices and their technical details are displayed (Figure 6.13). The assumed waste treatment routes and EWC codes are also provided for each plasterboard type (Figure 6.14).

Hence, this last component allows users to extract granular information about a material before performing the actual optimisation process. This data can also be exported to a spreadsheet to make further decisions.

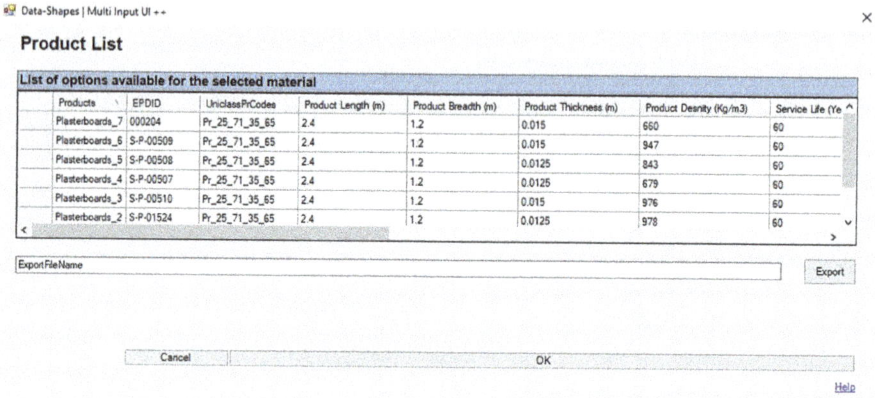

Data-Shapes | Multi Input UI ++

Product List

List of options available for the selected material

Products	EPDID	UniclassPrCodes	Product Length (m)	Product Breadth (m)	Product Thickness (m)	Product Desnity (Kg/m3)	Service Life (Ye
Plasterboards_7	000204	Pr_25_71_35_65	2.4	1.2	0.015	660	60
Plasterboards_6	S-P-00509	Pr_25_71_35_65	2.4	1.2	0.015	947	60
Plasterboards_5	S-P-00508	Pr_25_71_35_65	2.4	1.2	0.0125	843	60
Plasterboards_4	S-P-00507	Pr_25_71_35_65	2.4	1.2	0.0125	679	60
Plasterboards_3	S-P-00510	Pr_25_71_35_65	2.4	1.2	0.015	976	60
Plasterboards_2	S-P-01524	Pr_25_71_35_65	2.4	1.2	0.0125	978	60

ExportFileName

Export

Cancel

OK

Help

Figure 6.13 Technical Information of Plasterboard(s)

Source: Drafted by the author

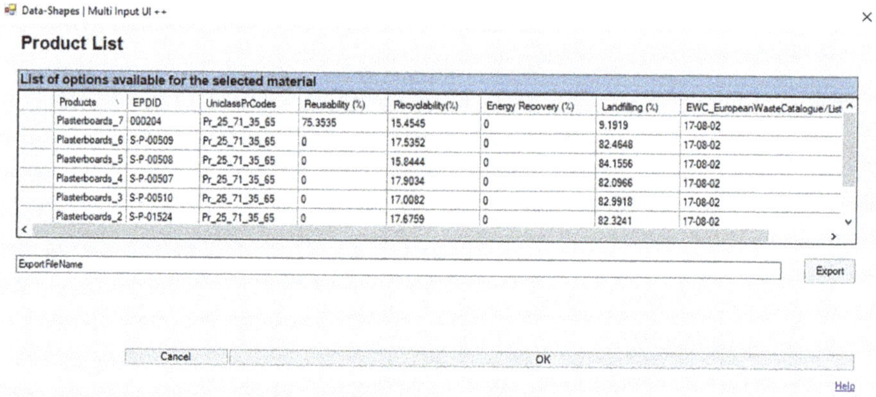

Data-Shapes | Multi Input UI ++

Product List

List of options available for the selected material

Products	EPDID	UniclassPrCodes	Reusability (%)	Recyclability(%)	Energy Recovery (%)	Landfilling (%)	EWC_EuropeanWasteCatalogue/List
Plasterboards_7	000204	Pr_25_71_35_65	75.3535	15.4545	0	9.1919	17-08-02
Plasterboards_6	S-P-00509	Pr_25_71_35_65	0	17.5352	0	82.4648	17-08-02
Plasterboards_5	S-P-00508	Pr_25_71_35_65	0	15.8444	0	84.1556	17-08-02
Plasterboards_4	S-P-00507	Pr_25_71_35_65	0	17.9034	0	82.0966	17-08-02
Plasterboards_3	S-P-00510	Pr_25_71_35_65	0	17.0082	0	82.9918	17-08-02
Plasterboards_2	S-P-01524	Pr_25_71_35_65	0	17.6759	0	82.3241	17-08-02

ExportFileName

Export

Cancel

OK

Help

Figure 6.14 Assumed Waste Treatment Routes of Plasterboard(s)

Source: Drafted by the author

6.4 BWT Validation

The case-study design helps researchers investigate 'how' and 'why' questions used to develop the overall context precisely [7]. Similarly, the final objective of the research is to identify how the BIM-integrated semantic framework is enhanced from the conventional CW measurement, reporting, and optimisation process in terms of interoperability and automation. In addition, this application-based evaluation is also considered appropriate for validating an ontology's vocabularies, concepts, data, hierarchies, and semantic relations [8]. It helps to decide the quality of the ontology based on the accuracy, consistency, and completeness of the results to

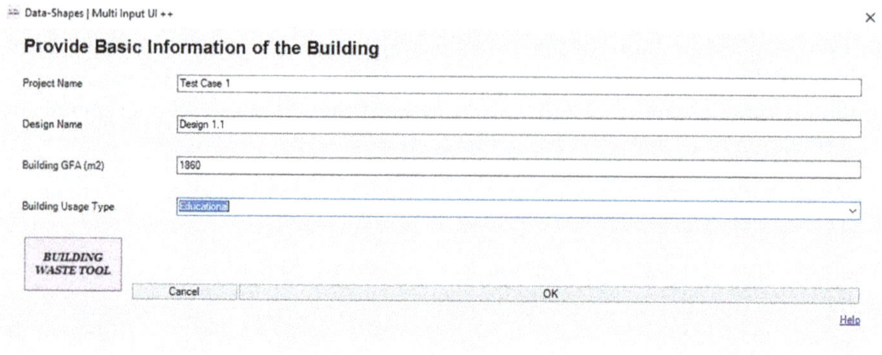

Figure 6.15 Basic Building and Design Information

Source: Drafted by the author

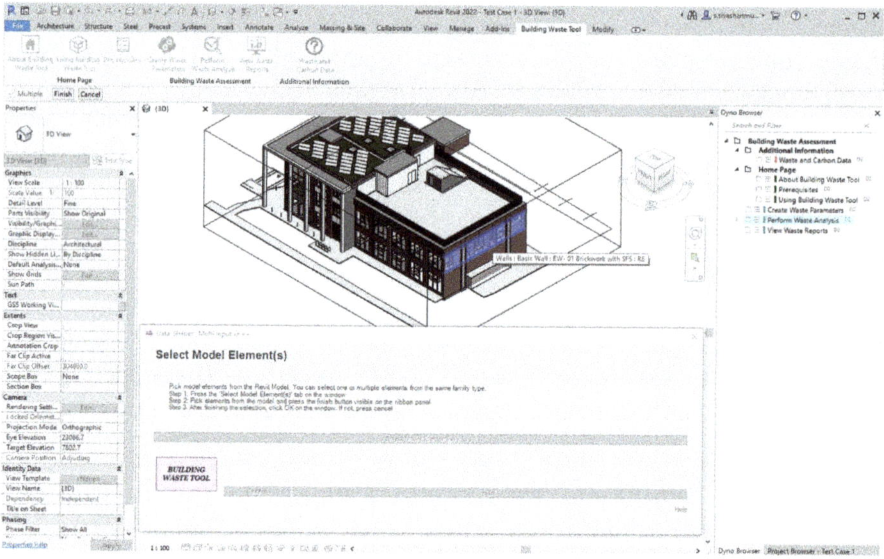

Figure 6.16 Wall Element Selected for Testing

Source: Drafted by the author

finish layer, metal studs as a structural layer, plasterboards as an inner finish layer, and non-rigid insulation and cavity as middle layers. Figure 6.17 depicts the material layers of a specific element within the selected family.

6.4.2.3 *Waste Information of Optimised Material(s)*

After element selection and material extraction, each material layer is analysed separately. The waste information of the material layers acts as custom parameters inside the project library (discussed in Section 6.3.2). Figure 6.18 presents the custom waste parameters of 1 kg of *AggregateBricks_3*.

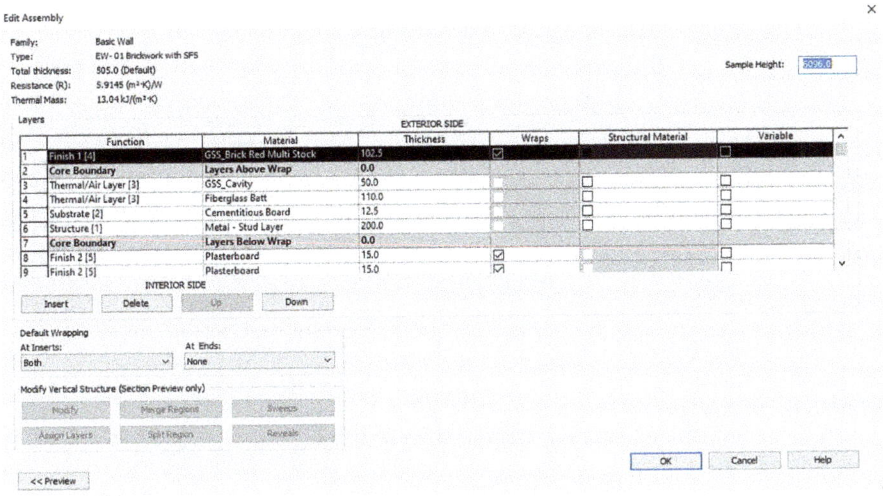

Figure 6.17 Material Layers of the Selected Wall Family Type

Source: Drafted by the author

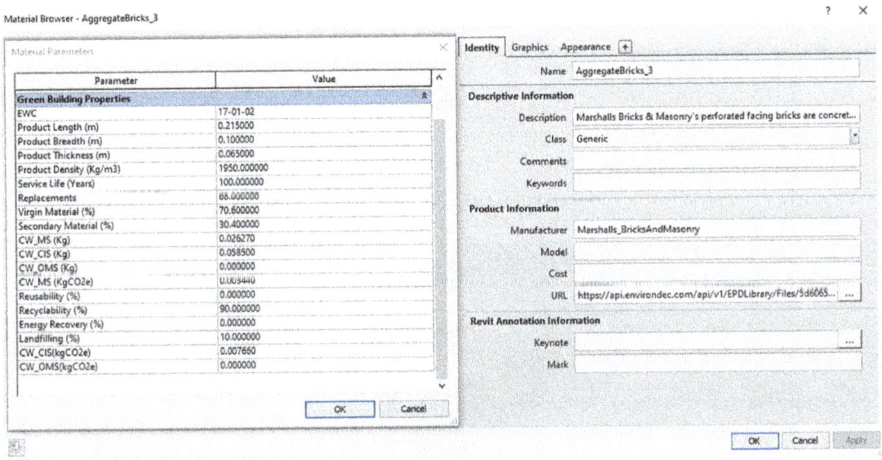

Figure 6.18 Custom Waste Parameters of AggregateBricks_3

Source: Drafted by the author

6.4.2.4 Optimised Material Layers of the Selected Element

The BWT runs through each material layer individually and replaces the existing materials with the optimised ones. The optimised material layers are applied to the duplicate family type 'EW-01 Brickwork with SFS_1' created before optimisation. Figure 6.19 presents the duplicate family type and the optimised material layers.

By running the analysis for a single element, other elements of the same family type are altered based on the user-selected list of elements. This family comprises ten wall elements (Figure 6.20), meaning they all share the same material layers.

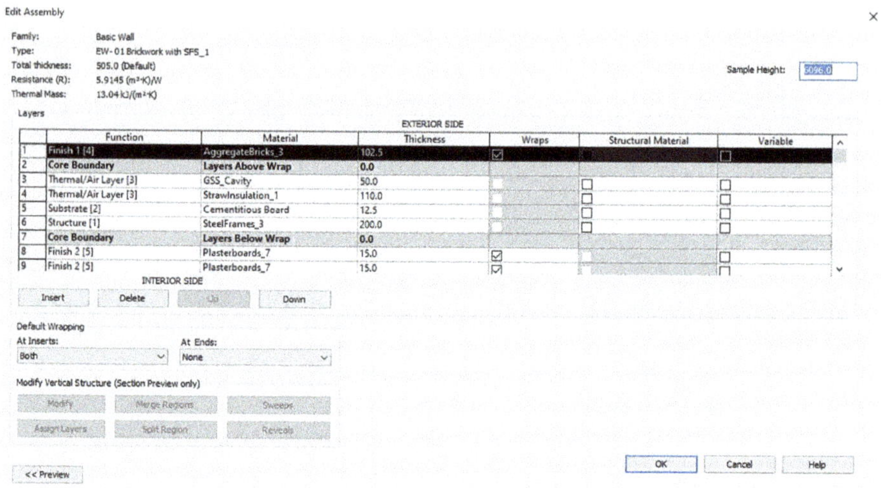

Figure 6.19 Optimised Material Layers of the Selected Wall Element

Source: Drafted by the author

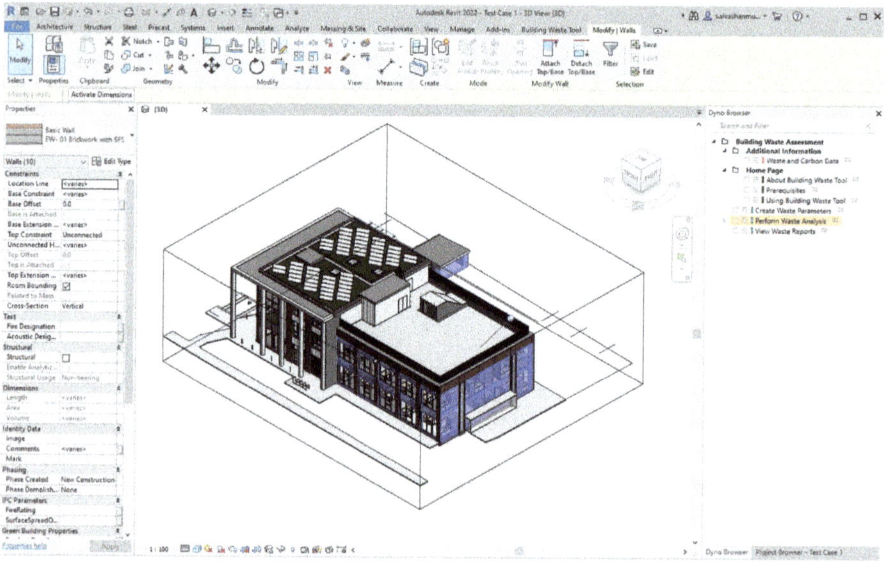

Figure 6.20 Number of Elements in the 'EW-01 Brickwork With SFS' Wall Family

Source: Drafted by the author

So, all the elements falling under the same family type are modified to align with optimisation results (Figure 6.21). In addition, the waste information of the corresponding element family could also be extracted as an MTO to facilitate communication between the project team and waste contractors.

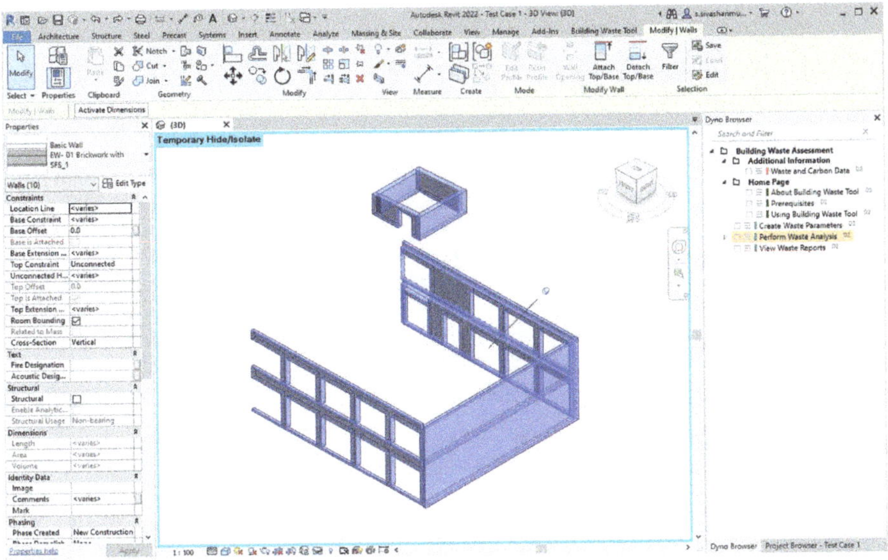

Figure 6.21 Optimising Multiple Elements of the Same Family Type

Source: Drafted by the author

The test-case operational results for a multi-layered external wall within the selected BIM model ensure that quantifying CW and informing optimal material choices using a BIM-integrated platform is seamless. Besides, the optimised materials are precisely mapped to the same material layer without disturbing the original element structure, thus obtaining the expected outcomes. This validates the first check that integrating the ontology and BIM achieves the core requirements expected from the CWQ framework: interoperability, standardisation, consistency, and automation.

6.4.3 *Test Case Quantification Results*

This section discusses the outcomes of the second check: evaluating the accuracy of the CW data. Since walls are the type of building elements incorporated into PRODCIRO, the material details within the walls are quantified and optimised. The test case building has 488 wall elements grouped into 23 wall families. Since the LOMD of the BIM model is <5, only the dimensional and basic material details are included in the model. The supply-chain details, including the product-specific names, manufacturer, and supplier details, are limited. Thus, the optimal material is selected from the list of available alternatives. Figure 6.22 shows a screenshot of a sample MTO of the wall category in the selected test case building.

The aggregated list of MTO showed that 19 different types of materials are used across the 488 elements. This includes '90 mm Timber Slat', 'Air Openings', 'Blockwork', 'Cementitious Board', 'Default Mesh Wall', 'Default Wall', 'Fiberglass Batt', 'GSS_Brick Red Multi Stock', 'GSS_Brick Black', 'GSS_Cavity', 'Metal – Stud Layer', 'Plasterboard', 'Render Material 255 - 255 - 255,' 'Roofing – EPDM Membrane', 'Soffit Cladding', 'Wall Cladding – Panels', 'Wall Cladding – Vertical 300 mm

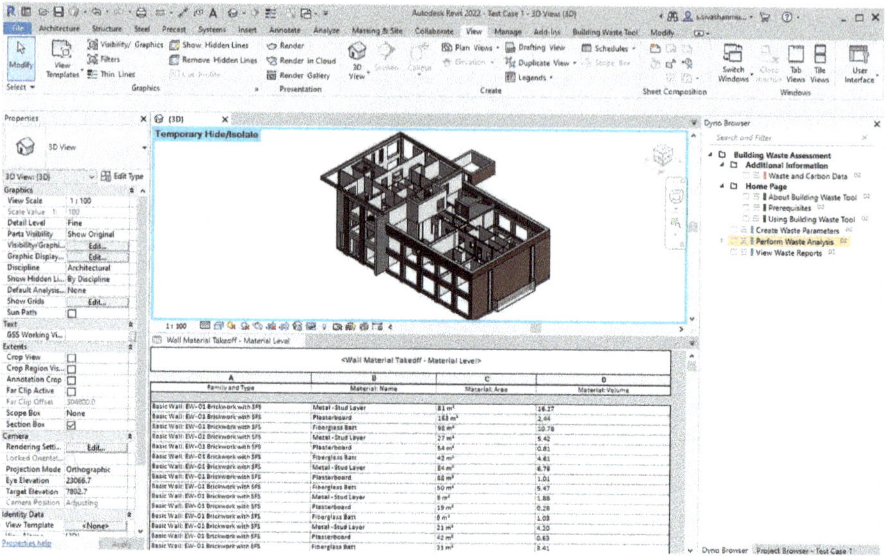

Figure 6.22 Screenshot of MTO of Wall Category in the Test Case Building

Source: Drafted by the author

Table 6.3 Materials Included for the Analysis

Material	*Volume (m^3)*	*CAWS*
'Metal – Stud Layer'	318.26	G10/150
'Plasterboard'	126.68	K10/165
'Fiberglass Batt'	116.49	F30/155
'Blockwork'	80.51	-
'GSS_Brick Red Multi Stock'	71.61	F10/110
'GSS_ Brick Black'	11.94	F10

Source: Drafted by the author

Panels', and 'Window Louvers'. Out of all, the filtered removed materials that had no specific Uniclass Pr codes. For instance, the 'Default Wall' material lacks a specific material composition to allow the tool to quantify CW. So, the tool removes these materials.

Furthermore, materials like cementitious boards and cladding panels had either no or no single EPDs that met the geographical scope criteria. Hence, the lack of data limits the inclusion of these materials in the CWQ process. Finally, the tool filters the six material types that met the prerequisite conditions. Table 6.3 shows a list of materials that are included in the analysis. The CAWS codes and matched Uniclass Pr codes are also included.

Since geographical location plays a vital role in deciding the CW generation levels, the alternatives are only products with the UK as their geographical scope. The geographical scope of products is verified using EPD certificates. The design details, technical factors influencing waste generation, and waste generation levels of optimised materials are tabulated in Table 6.4. The obtained values are the impacts of various CW indicators like mass, density, service life, reused and

Table 6.4 Optimal Material Choices and Their CW Values

Existing materials	*Volume (m^3)*	*Optimised materials*	*Uniclass Pr codes*	*Product density (Kg/m^3)*	*CW_MS (tonnes)*	*CW_MS (tCO2e)*[3]	*CW_CIS (tonnes)*	*CW_CIS (tCO2e)*[3]	*Service life (years)*	*Replacements*	*CW_OMS (tonnes)*	*CW_OMS (tCO2e)3*
'Metal-Stud Layer'	318.26	*SteelFrames_3*	Pr_20_85_32_84	7750	29.598	5.03E+0	200.281	3.40E+01	60	0	0.00	0.00E+0
'Plasterboard'	126.68	*Plasterboards_7*	Pr_25_71_35_65	560	1.166	1.05E-02	4.013	3.62E-02	60	0	0.00	0.00E+0
'Fiberglass Batt'	116.49	*StrawInsulation_1*	Pr_25_57_06	100	0.126	4.37E-05	0.228	7.92E-05	75	0	0.00	0.00E+0
'Blockwork[1]'	80.51	*ConcreteBlocks_2*	Pr_20_93_52_05	550	10.73	2.96E+01	3.061	8.45E+00	100	0	0.00	0.00E+0
'GSS_Brick Red Multi Stock'	71.61	*AggregateBricks_3*	Pr_20_93_52_02	1950	3.672	4.08E-02	8.169	9.07E-02	100	0	0.00	0.00E+0
'GSS_Brick Black[2]'	11.94	*AggregateBricks_3*	Pr_20_93_52_02	1950	0.658	8.24E-03	1.362	1.71E-02	100	0	0.00	0.00E+0

Acronyms: CW_MS – Construction Waste generated during the product Manufacturing Stage; CW_CIS – Construction Waste generation during Construction/Installation Stages; and CW_OMS – Construction Waste generation during Operation and Maintenance Stages.

Source: Drafted by the author

Notes:

1 The term 'blockwork' in the Uniclass is classified under the 'Systems' category, which does not align with the classification supported by PRODCIRO. So, the tool accounts for the primary material concrete block used to construct blockwork.

2 The bricks are currently classified based on their material composition in the Uniclass Pr classification. Due to the lack of colour factors in this classification, the optimisation approach adopted for GSS_Brick Black is similar to GSS_Brick Red Multi Stock. The results may slightly differ if the GWP values for black bricks are different from red-coloured bricks.

3 The scientific expressions in CW__MS (tCO2e), CW__CIS (tCO2e), and CW__OMS (tCO2e) are used for communication purposes. The 'number' format was used to embed the CW impact data in the BIM model (as depicted in Figure 6.18).

recycled materials percentage, and embodied carbon associated with the material manufacturing process.

Accounting for weight-based results across the product manufacturing (CW_MS), construction (CW_CIS), and operation and maintenance stages (CW_OMS), the *SteelFrames_3* generates a high amount of waste followed by *AggregateBricks_3*, *ConcreteBlocks_2*, *Plasterboards_7*, and *StrawInsualtion_1*. The *SteelFrames_3* is predicted to generate 229.88 tonnes of waste during the manufacturing and construction/installation stages, while the waste generation during the operation and maintenance stages is zero because of no expected replacements. The *SteelFrames_3*, being a structural material, has its service life equivalent to the assumed RSP of the building, thereby leading to zero waste generation during the operation and maintenance stages.

The *AggregateBricks_3* is filtered as an optimal choice for 'GSS_Brick Red Multi Stock' and 'GSS_ Brick Black'. The CW_MS and CW_CIS of the optimised material is predicted to generate 14.133 tonnes. Similar to *SteelFrames_3*, the waste generated during the operation and maintenance stages of *AggregateBricks_3* is zero. Since the service life of *AggregateBricks_3* (100 years) is higher than the assumed RSP of the building (60 years), it is assumed that material will only be removed during the EoL stage, so there is no expected waste generation in the operation and maintenance stages (discussed in Section 6.3.3). This is also the scenario with *ConcreteBlocks_2*, which holds a service life of 100 years. However, their waste generation during the manufacturing and construction stages is considerable, contributing to about 13.791 tonnes. The least waste-contributing materials accounting for their weight are *Plasterboards_7* and *StrawInsulation_1*. The CW_MS and CW_CIS of the former and latter are 5.179 and 0.354 tonnes, respectively. Like other materials, these materials generate no waste during operation and maintenance.

The report by [9] highlighted that the waste materials dominating the list would differ based on the selected estimation type. It is emphasised that inert waste will dominate weight-based waste estimations, and insulation and packaging materials will become more dominant if waste volumes are accounted for in the analysis. Thirdly, these two trends are predicted to differ if the environmental impacts of the product are incorporated into quantification [9]. Other research studies [10–15] also concluded that inert materials like bricks, blocks, and concrete generate large amounts of waste compared to other material types by accounting for their weight. However, this is based on estimating the whole building. Since only a specific element category is selected to test the workflow, the results cannot be directly compared to draw conclusions. In our case, the selected building is a steel frame, so the quantity of steel frames is higher than other materials. So, because of higher material quantity, the quantity of metallic material (*SteelFrames_3*) waste dominated both weight- and impact-based CW estimations.

Table 6.5 presents the usage of secondary materials, predicted waste treatment scenarios, and EWC of optimised materials. The results show that 83% of *SteelFrames_3* production used secondary materials, contributing to higher circularity levels. Followed by *SteelFrames_3* is *AggregateBricks_3*, where 32% of its

Table 6.5 Optimal Material Choices, Resource Use, and Waste Treatment Routes

Optimised materials	*Secondary material (%)*	*Reusability (%)*	*Recyclability (%)*	*Energy recovery (%)*	*Landfilling (%)*	*EWC*
SteelFrames_3	83	0	92	0	8	17-04-05
Plasterboards_7	4	75	15	0	10	17-08-02
StrawInsulation_1	0	0	0	95	5	17-06-04
ConcreteBlocks_2	NA	NA	NA	NA	NA	17-01-01
AggregateBricks_3	32	0	90	0	10	17-01-02
AggregateBricks_3	32	0	90	0	10	17-01-02

Source: Drafted by the author

composition is made of secondary materials. Although this is just one-third of the total material quantity, the change represents implementing CE principles to reuse materials in the built environment. Other materials like *Plasterboards_7* and *StrawInsulation_1* primarily used virgin materials for manufacturing. However, it is vital to note that straw bales are sustainable products in nature, and so despite the zero use of secondary material, the *StrawInsulation_1* emphasise circularity and supports carbon reduction.

Besides, the treatment routes for each CW are unique. Considering the principles of Waste Hierarchy and CE principles, reusing materials without further treatment should prioritised if waste prevention at source is unachievable. Furthermore, followed by reuse are recycling, energy recovery, and landfilling (discussed in Section 2.3.3). However, the quantification results show that recycling looks dominant over other waste treatment routes. Nine-tenths of *SteelFrames_3* and *AggregateBricks_3* waste is expected to undergo recycling. Moreover, the remaining one-tenth of the waste is expected to be landfilled. In the case of *Plasterboards_7*, 75% is expected to be reused, 15% recycled, and 10% landfilled. Besides, 95% of *StrawInsulation_1 waste* is recovered as energy, and only 5% is expected to undergo landfilling. Although it is widely discussed in the literature that concrete blocks have reasonable recycling rates, there is insufficient data regarding this in the filtered EPDs for *ConcreteBlocks_2*.

6.4.4 ***Comparison of CW Values of Optimised Materials With Other Datasets***

The obtained CW values are compared with the waste data of existing databases to validate the accuracy of the results. The current research quantifies the CW generation and its impacts across the project life cycle. However, no databases provide a sufficient waste range for all these stages. So, based on the thorough analysis of existing CW reporting tools in the UK and the research studies, it is evident that only a few survey reports and databases provide a wastage range for building materials. In addition, these databases only focus on the construction/installation stages. Besides, the extraction of actual waste data generated on-site is at the material level for the whole building and lacks classification based on elements. So, utilising this recorded data will be inappropriate as the PRODCIRO entities currently focus on wall elements. So, accounting for the data limitations, the wastage levels of materials during the construction and installation stages are validated against survey reports and on-site waste-data collection tools. In addition, the product loss expected by the manufacturer during the construction/installation of the product is also included in the validation process. Table 6.6 shows the estimated wastage values of materials during their construction/installation stage against the existing waste-data reports.

Out of six materials, only two have sufficient data from the manufacturers to compare the accuracy of estimated CW values. The estimated wastage level of *SteelFrames_3* is 8% of the total material quantity, which is 3% higher than the expected product waste by the manufacturer. Secondly, the waste percentage of *Plasterboards_7* is 6%, 4% lower than the expected product loss. Although there is no waste data for *ConcreteBlocks_2* from the manufacturer, the survey conducted by Reusefully Ltd states that the wastage rate of concrete blocks ranges from 3% to 5% [16]. Considering this range, the estimated CW for *ConcreteBlocks_2* is 1% higher than the prescribed range. This increased CW in *SteelFrames_3,* and *ConcreteBlocks_2* might be a result of unstandardised designs leading to increased offcuts (discussed in Section 4.4.5.1), improper handling of materials, and not utilising the supplier's take-back schemes [16]. Thus, the project should incorporate these factors to reduce the CW generation levels. In many cases, the products are designed to be in custom sizes as required for specific designs, which, when used, will prevent the waste generated during installation on-site. However, the wastage levels of all materials are approximately close to the 'good' wastage rate proposed by the WRAP [17].

Overall, the CW quantification results from the research study will provide the accurate and consistent data required for developing comprehensive WM from the early

Table 6.6 Comparison of Estimated Wastage With Existing CW Data Reports

		Existing CW data sources			
Optimised materials	*Estimated wastage_ CIS*	*Expected product loss assumed by the manufacturer*	*Reusefully Ltd.'s survey report*	*WRAP wastage rate (Good)**	*WRAP wastage rate (Baseline)**
SteelFrames_3	8%	5%	NA	5%	15%
Plasterboards_7	6%	10%	NA	15%	22.5%
StrawInsulation_1	3%	Assumed as 0%	NA	NA	NA
ConcreteBlocks_2	6%	NA	3 (Good rate) to 5 (Baseline rate)	5%	20%
AggregateBricks_3	6%	NA	NA	5%	20%
AggregateBricks_3	6%	NA	NA	5%	20%

* The latest report was published in 2008, so this should be accounted for while making decisions.

Source: Drafted by the author

design stages. At present, the ontology primarily focuses on integrating the indicators at the material and product level to quantify and optimise CW generation at the early stages of the project. Furthermore, in addition to product-level factors, the design and construction factors that influence the CW generation levels could be added to the ontology to facilitate a comprehensive CWQ process at the building level. This integration further allows the user to optimise the processes and activities carried out across the project life cycle to maximise the CW reduction potentials of buildings.

6.5 Summary

This chapter presented the components of BWT developed to quantify, classify, and optimise CW from the early design stages. This chapter also critically discussed the information standards, visualisation, and API functionalities facilitated by BIM and SWT in developing a UI. The integration of BIM and PRODCIRO enabled seamless integration of design, material, CW, and life cycle impact data. Besides, the application of the whole lifecycle carbon assessment assisted the framework to support granular and accurate CW quantification across product manufacturing, construction/installation, and operation and maintenance stages. The CW results showed that the material quantity decides the quantity of CW generated. Since the selected test case building is a steel frame building, the use of steel frames is high, thus contributing to a large volume of metallic waste generation. So, proposing designs that consume fewer materials would significantly impact CW reduction, and incorporating CE principles would minimise the impacts caused by excessive CW generation and treatment. The results also showed that recycling is widely applied for inert and metallic waste materials, while other waste, like insulation and plasterboard, undergoes minimal recycling.

References

1. S. Sivashanmugam, S. Rodriguez, F.P. Rahimian, F. Elghaish and N. Dawood, Enhancing information standards for automated construction waste quantification and classification. *Automation in Construction*, 2023. 152: p. 104898. https://doi.org/10.1016/j.autcon.2023.104898
2. A. Talla and S. McIlwaine, Industry 4.0 and the circular economy: Using design-stage digital technology to reduce construction waste. *Smart and Sustainable Built Environment*, 2024. 13(1): pp. 179–198. https://doi.org/10.1108/SASBE-03-2022-0050
3. O. Disney, M. Roupé, M. Johansson and A.D. Leto, Embracing BIM in its totality: A total BIM case study. *Smart and Sustainable Built Environment*, 2022. 13(3): pp. 512–531. https://doi.org/10.1108/SASBE-06-2022-0124
4. A. Kamari, B.M. Kotula and C.P.L. Schultz, A BIM-based LCA tool for sustainable building design during the early design stage. *Smart and Sustainable Built Environment*, 2022. 11(2): pp. 217–244. https://doi.org/10.1108/SASBE-09-2021-0157
5. S. Sivashanmugam, R. Sergio, R. Farzad and D. Nashwan, Maximising the construction waste reduction potential – Barriers and catalysts. *Proceedings of the Institution of Civil Engineers – Civil Engineering*, 2022. 176(6): pp. 1–9. https://doi.org/10.1680/jcien.22.00163
6. RICS, *Whole life carbon assessment for the built environment*, 2023. https://www.rics.org/profession-standards/rics-standards-and-guidance/sector-standards/construction-standards/whole-life-carbon-assessment. Last access: 08 December 2024.

7. R.K. Yin, *Case study research: Design and methods*, Vol. 5, 2009. Sage. ISBN: 1412960991.
8. J. Raad and C. Cruz, A survey on ontology evaluation methods, in *Proceedings of the international conference on knowledge engineering and ontology development, part of the 7th international joint conference on knowledge discovery, knowledge engineering and knowledge management*, 2015. https://doi.org/10.5220/0005591001790186
9. CLC, *Zero avoidable waste in construction*, 2020. https://www.constructionleadershipcouncil.co.uk/wp-content/uploads/2016/05/ZAW-Report-Final-Draft-25-February–2020.pdf. Last access: 15 May 2023.
10. A. Bakshan, I. Srour, G. Chehab and M. El-Fadel, A field based methodology for estimating waste generation rates at various stages of construction projects. *Resources, Conservation and Recycling*, 2015. 100: pp. 70–80. https://doi.org/10.1016/j.resconrec.2015.04.002
11. F. Kleemann, H. Lehner, A. Szczypińska, J. Lederer and J. Fellner, Using change detection data to assess amount and composition of demolition waste from buildings in Vienna. *Resources, Conservation and Recycling*, 2017. 123: pp. 37–46. https://doi.org/10.1016/j.resconrec.2016.06.010
12. J. Li, Z. Ding, X. Mi and J. Wang, A model for estimating construction waste generation index for building project in China. *Resources, Conservation and Recycling*, 2013. 74: pp. 20–26. https://doi.org/10.1016/j.resconrec.2013.02.015
13. P. Sun, N. Zhang, J. Zuo, R. Mao, X. Gao and H. Duan, Characterizing the generation and flows of building interior decoration and renovation waste: A case study in Shenzhen City. *Journal of Cleaner Production*, 2020. 260: p. 121077. https://doi.org/10.1016/j.jclepro.2020.121077
14. A. Vilventhan, V. Ram and S. Sugumaran, Value stream mapping for identification and assessment of material waste in construction: A case study. *Waste Management & Research*, 2019. 37(8): pp. 815–825. https://doi.org/10.1177/0734242X19855429
15. H. Wu, H. Duan, L. Zheng, J. Wang, Y. Niu and G. Zhang, Demolition waste generation and recycling potentials in a rapidly developing flagship megacity of South China: Prospective scenarios and implications. *Construction and Building Materials*, 2016. 113: pp. 1007–1016. https://doi.org/10.1016/j.conbuildmat.2016.03.130
16. G.H. Katherine Adams, *Final report: Wastage rates for blocks and ready mixed concrete*, 2023. https://www.mpamasonry.org/MPAMasonry/media/root/assets/MPA-Wastage-Desk-Final-report-01-06-23v3.pdf. Last access: 15 August 2024.
17 .Emma Burton and Nick Friedrich, *Net waste tool – Guidance to reference data, version 1.0*, 2008. https://www.eugris.info/displayresource.aspx?r=6438&Cat=software. Last access: 10 August 2024.

Appendices

Below is the list of alternatives for the materials selected for CW assessment using SPARQL queries. The alternative choices are those filtered after the stage 1 filter, that is, based on its 'applicableForBuildingElements' property (discussed in Section 6.3.3).

Appendix A.1

Alternative Material Choices for Metal Studs

Table 6.7 Alternative Material Choices for Metal Studs Along with their CW Indicator Results

The estimated results are for the 318.26 m³ of metal studs (the estimated volume from the BIM model – Table 6.3).

Material choices	*EPD ID*	*Uniclass Pr codes*	*Product density (Kg/m³)*	*CW_MS (tonnes)*	*CW_MS (tCO2e)*	*CW_CIS (tonnes)*	*CW_CIS (tCO2e)*	*Service life (years)*	*Replacements*	*CW_OMS (tonnes)*	*CW_OMS (tCO2e)*	*SM (%)*	*Reusability (%)*	*Recyclability (%)*	*Energy recovery (%)*	*Landfilling (%)*	*EWC*
SteelFrames_3	S-P-12739	Pr_20_85_32_84	7750	29.598	5.03E+00	200.281	3.40E+01	60	0	0.00	0.00	83	0	92	0	8	17-04-05
SteelFrames_5	HUB-0090		7850	269.309	5.21E+01	202.865	3.93E+01	60	0	0.00	0.00	95.2	0	99	0		
SteelFrames_4	HUB-0089		7850	269.309	1.54E+02	202.865	1.16E+02	60	0	0.00	0.00	0.01	0	99	0	1	
SteelFrames_1	S-P-12275		7750	12307.91	2.02E+05	200.28	3.29E+03	60	0	0.00	0.00	13.90	0	92	0	8	
SteelFrames_2	HUB-0024		7850	23589.34	4.49E+05	202.865	3.86E+03	50	0	0.2	0.00*	0.00*	0	95	0	5	

Acronyms: CW_MS – Construction Waste generated during Product Manufacturing Stages; CW_CIS – Waste generation during Construction/Installation Stages; CW_OMS – Construction Waste generation during Operation and Maintenance Stages; SM – Secondary Material; EWC – European Waste Catalogue; and 'NA' denotes that the specific data is 'not available'.

Notes:

* Since the replacement is less than 1, quantifying CW will not be appropriate. Instead, extending the service life of the material through additional maintenance would be a preferred option.

Appendix A.2

Alternative Material Choices for Plasterboard

Table 6.8 Alternative Material Choices for Plasterboard Along with their CW Indicator Results

The estimated results are for the 126.68 m³ of plasterboard (the estimated volume from the BIM model – Table 6.3).

Material choices	*EPD ID*	*Uniclass Pr codes*	*Product density (Kg/m³)*	*CW_MS (tonnes)*	*CW_MS (tCO2e)*	*CW_CIS (tonnes)*	*CW_CIS (tCO2e)*	*Service life (years)*	*Replacements*	*CW_OMS (tonnes)*	*CW_OMS (tCO2e)*	*SM (%)*	*Reusability (%)*	*Recyclability (%)*	*Energy recovery (%)*	*Landfilling (%)*	*EWC*
Plasterboards_7	000204	Pr_25_71_35_65	670	1.116	1.05E-02	4.013	3.62E-02	60	0	0.00	0.00	4.03	75.35	15.45	0	9.19	17-08-02
Plasterboards_3	S-P-00510		976	0.281	2.92E-03	5.935	6.17E-02	60	0	0.00	0.00	1.06	0	17.01	0	82.99	
Plasterboards_4	S-P-00507		679	0.112	1.19E-03	4.129	4.41E-02	60	0	0.00	0.00	0.97	0	17.90	0	82.10	
Plasterboards_1	S-P-00506		675	0.169	1.91E-03	4.104	4.62E-02	60	0	0.00	0.00	4.08	0	18.10	0	81.90	
Plasterboards_6	S-P-00509		947	0.287	3.23E-03	5.758	6.48E-02	60	0	0.00	0.00	6.18	0	17.54	0	82.46	
Plasterboards_2	S-P-01524		978	0.152	1.96E-03	5.947	7.67E-02	60	0	0.00	0.00	2.60	0	17.68	0	82.32	
Plasterboards_5	S-P-00508		843	5.011	1.18E-01	5.126	1.21E-01	60	0	0.00	0.00	15.28	0	15.84	0	84.16	

Acronyms: CW_MS – Construction Waste generated during Product Manufacturing Stages; CW_CIS – Waste generation during Construction/Installation Stages; CW_OMS – Construction Waste generation during Operation and Maintenance Stages; SM – Secondary Material; EWC – European Waste Catalogue; 'NA' denotes that the specific data is 'not available'.

Appendix A.3

Alternative Material Choices for Fiberglass Batt

Table 6.9 Alternative Material Choices for Fiberglass Batt Along with their CW Indicator Results

The estimated results are for 116.49 m^3 of fibreglass batt (the estimated volume from the BIM model; Table 6.3).

Material choices	*EPD ID*	*Uniclass Pr codes*	*Product density (Kg/m³)*	*CW_MS (tonnes)*	*CW_MS (tCO2e)*	*CW_CIS (tonnes)*	*CW_CIS (tCO2e)*	*Service life (years)*	*Replacements*	*CW_OMS (tonnes)*	*CW_OMS (tCO2e)*	*SM (%)*	*Reusability (%)*	*Recyclability (%)*	*Energy recovery (%)*	*Landfilling (%)*	*EWC*
StrawInsulation_1	S-P-03854	Pr_25_57_06	100	0.126	4.37E-05	0.228	7.92E-05	75	0	0.00	0.00	0	0	0	95	5	17-06-04
GlassWoolInsulation_1	000371	Pr_25_57_06_34	22	0.736	3.04E-01	0.0502	2.08E-02	60	0	0.00	0.00	100	0	0	0	100	
GlassWoolInsulation_2	000372	Pr_25_57_06_34	20	0.478	1.06E+00	0.046	1.53E-01	60	0	0.00	0.00	100	0	0	0	100	

Acronyms: CW_MS – Construction Waste generated during Product Manufacturing Stages; CW_CIS – Waste generation during Construction/Installation Stages; CW_OMS – Construction Waste generation during Operation and Maintenance Stages; SM – Secondary Material; EWC – European Waste Catalogue; and 'NA' denotes that the specific data is 'not available'.

Appendix A.4

Alternative Material Choices for Blockwork*

Table 6.10 Alternative Material Choices for Blockwork Along with their CW Indicator Results

The estimated results are for the 80.51 m³ of concrete blocks (the estimated volume from the BIM model; Table 6.3).

Material choices	*EPD ID*	*Uniclass Pr codes*	*Product density (Kg/m³)*	*CW_MS (tonnes)*	*CW_MS (tCO2e)*	*CW_CIS (tonnes)*	*CW_CIS (tCO2e)*	*Service life (years)*	*Replacements*	*CW_OMS (tonnes)*	*CW_OMS (tCO2e)*	*SM (%)*	*Reusability (%)*	*Recyclability (%)*	*Energy recovery (%)*	*Landfilling (%)*	*EWC*
ConcreteBlocks_2	EPDIE-18–10	Pr_20_93_52_05	650	10.73	2.96E+01	3.061	8.45E+00	100	0	0.00	0.00	NA	NA	NA	NA	NA	17-01-01
ConcreteBlocks_3	EPDIE-18–10		760	13.40	3.79E+01	3.579	1.01E+01	100	0	0.00	0.00	NA	NA	NA	NA	NA	
ConcreteBlocks_1	EPDIE-18–10		480	9.16	2.77E+01	2.261	6.83E+00	100	0	0.00	0.00	NA	NA	NA	NA	NA	

Acronyms: CW_MS – Construction Waste generated during Product Manufacturing Stages; CW_CIS – Waste generation during Construction/Installation Stages; CW_OMS – Construction Waste generation during Operation and Maintenance Stages; SM – Secondary Material; EWC – European Waste Catalogue; and 'NA' denotes that the specific data is 'not available'.

Notes:

* The term 'blockwork' in the Uniclass is classified under the 'Systems' category, which does not align with the classification supported by PRODCIRO. So, the tool accounts for the primary material concrete block used to construct blockwork.

Appendix A.5

Alternative Material Choices for GSS_Brick Red Multi Stock

Table 6.11 Alternative Material Choices for GSS_Brick Red Multi Stock Along with their CW Indicator Results

The estimated results are for the 71.61 m³ of bricks (the estimated volume from the BIM model – Table 6.3).

Material choices	*EPD ID*	*Uniclass Pr codes*	*Product density (Kg/m³)*	*CW_MS (tonnes)*	*CW_MS (tCO2e)*	*CW_CIS (tonnes)*	*CW_CIS (tCO2e)*	*Service life (years)*	*Replacements*	*CW_OMS (tonnes)*	*CW_OMS (tCO2e)*	*SM (%)*	*Reusability (%)*	*Recyclability (%)*	*Energy recovery (%)*	*Landfilling (%)*	*EWC*
ReclaimedBricks_1	00503	Pr_20_93_52_15	1650	1.465	2.26E-03	8.956	1.38E-02	50	0.2	0.00*	0.00*	100	0	99	0	1	17-01-02
AggregateBricks_3	S-P-03536	Pr_20_93_52_02	1950	3.672	4.08E-02	8.169	9.07E-02	100	0	0.00	0.00	30.40	0	90	0	10	17-01-01
AggregateBricks_2	S-P-03536	Pr_20_93_52_02	1950	3.944	4.94E-02	8.169	1.02E-01	100	0	0.00	0.00	31.66	0	90	0	10	17-01-01
AggregateBricks_1	S-P-03536	Pr_20_93_52_02	1950	3.733	5.93E-02	8.169	1.30E-01	100	0	0.00	0.00	31.12	0	90	0	10	17-01-01
ClayBricks_1	00092	Pr_20_93_52_15	1900	0.926	1.63E-02	10.313	1.81E-01	150	0	0.00	0.00	0	0	90	0	10	17-01-02

Acronyms: CW_MS – Construction Waste generated during Product Manufacturing Stages; CW_CIS – Waste generation during Construction/Installation Stages; CW_OMS – Construction Waste generation during Operation and Maintenance Stages; SM – Secondary Material; EWC – European Waste Catalogue; and 'NA' denotes that the specific data is 'not available'.

Notes:

* Since the replacement is less than 1, quantifying CW will not be appropriate. Instead, extending the service life of the material through additional maintenance

Appendix A.6

Alternative Material Choices for GSS_ Brick Black[1]

Table 6.12 Alternative Material Choices for GSS_Brick Black Along with their CW Indicator Results

The estimated results are for the 11.94 m³ of bricks (the estimated volume from the BIM model – Table 6.3).

Material choices	*EPD ID*	*Uniclass Pr codes*	*Product density (Kg/m³)*	*CW_MS (tonnes)*	*CW_MS (tCO2e)*	*CW_CIS (tonnes)*	*CW_CIS (tCO2e)*	*Service life (years)*	*Replacements*	*CW_OMS (tonnes)*	*CW_OMS (tCO2e)*	*SM (%)*	*Reusability (%)*	*Recyclability (%)*	*Energy recovery (%)*	*Landfilling (%)*	*EWC*
ReclaimedBricks_1	00503	Pr_20_93_52_71	1650	0.244	3.77E-04	1.493	2.30E+00	50	0.2	0.00[2]	0.00[2]	100	0	99	0	1	17-01-02
AggregateBricks_3	S-P-03536	Pr_20_93_52_02	1950	0.612	6.80E-03	1.362	1.51E-02	100	0	0.00	0.00	30.40	0	90	0	10	17-01-01
AggregateBricks_2	S-P-03536	Pr_20_93_52_02	1950	0.658	8.24E–0.3	1.362	1.71E-02	100	0	0.00	0.00	31.66	0	90	0	10	17-01-01
AggregateBricks_1	S-P-03536	Pr_20_93_52_02	1950	0.622	9.89E-03	1.362	2.17E-02	100	0	0.00	0.00	31.12	0	90	0	10	17-01-01
ClayBricks_1	00092	Pr_20_93_52_15	1900	0.154	2.72E-03	1.719	3.03E-02	150	0	0.00	0.00	0	0	90	0	10	17-01-02

Acronyms: CW_MS – Construction Waste generated during Product Manufacturing Stages; CW_CIS – Waste generation during Construction/Installation Stages; CW_OMS – Construction Waste generation during Operation and Maintenance Stages; SM – Secondary Material; EWC – European Waste Catalogue; and 'NA' denotes that the specific data is 'not available'.

Notes:

1 The bricks are currently classified on the basis of their material composition in the Uniclass Pr classification. Due to the lack of colour factors in this classification, the optimisation approach adopted for GSS_Brick Black is similar to GSS_Brick Red Multi Stock. The results may slightly differ if the GWP values for black bricks are different from red-coloured bricks.

2 Since the replacement is less than 1, it will not be appropriate to conduct a CW assessment. Instead, extending the service life of the material through additional maintenance would be a preferred option.

7 Potentialities of the Construction Waste Quantification Framework

7.1 Introduction

The ununified data structure, non-standardised information, and inconsistent recording of waste data were some of the shortcomings identified in the existing CWQ models. So, the study explored 'what type of information' and 'how the data should be structured and standardised' to ensure a linear, interoperable, and automated workflow. The outputs from the literature review and thematic and exploratory data analysis are integrated to develop a BIM-integrated semantic framework that enhances interoperability and automation across the CWQ process. The ontology (prefixed as PRODCIRO) is used to demonstrate the application of the proposed framework. The PRODCIRO proposes a solution that links CW data from multiple sources and defines semantic relations between the CW entities. It also aids in inferring logical and meaningful insights based on established relations. Besides, to reduce the level of human intervention across the CWQ workflow and ensure the proposed framework complies with BIM, a user-interactive scalable tool (named BWT) is introduced at the proof-of-concept level. The framework is also validated against a real-life test case building. In addition to achieving a unified data structure and standardised outputs, the integration of BIM and SWT yielded comprehensive results. The potentialities of the proposed workflow obtained during the design, development, implementation, and validation of the CWQ framework are discussed in this chapter.

7.2 Potentialities of the Workflow Proposed in the BIM-Integrated Semantic CWQ Framework

This section discusses the potentialities of the CWQ workflow proposed in the BIM-integrated semantic framework. It includes the following: (i) a granular set of CW indicators promise quantification and optimisation accuracy and transparency, (ii) standardised CW data ensure interoperability and automation, (iii) diverse CW measurement units ensure integrated decision-making, (iv) the Uniclass classification system enhance the material optimisation choices, (v) the semantic relationships established using an ontology unify the difference between material and CW

DOI: 10.1201/9781003610762-7

characterisation codes, and (vi) the proposed framework would act an integrated CW data hub supporting BIM Level 3 adoption.

7.2.1 A Granular Set of CW Indicators Promise Quantification and Optimisation Accuracy and Transparency

In the current scenario, WGR or waste factors of materials are the primary indicators employed for quantifying and optimising CW. Even the most advanced ICT-integrated models use only the material volume and WGRs to quantify the CW generation (discussed in Sections 2.7.4 and 4.4.5.3) without considering other influencing factors. However, the results from the literature and exploratory data analysis proved that technical factors, including density, service life, dimensions, and building usage, play a vital role in CW-informed decision-making. For instance, the test case operational results show that the *AggregateBricks_3* is informed as an optimal alternative for 'GSS_Brick Red Multi Stock' and 'GSS_ Brick Black' (discussed in Section 6.4.3). Although the CW results across the manufacturing, construction/installation, and operation and maintenance stages show that *ReclaimedBricks_1* has less impact than other materials, it is not selected by the tool as an optimal choice. This is because of the low service life of *ReclaimedBricks_1* compared to *AggregateBricks_3*. The service of the former was 50 years, which is less than the service life of the latter (60 years). The CW indicators of alternative choices for bricks are tabulated in the Appendices provided in Chapter 6. Since the wall is a structural element, the service life of structural materials should be equal to the service life of the building. So, the optimisation results prove that the technical functions should be prioritised in addition to environmental factors while optimising materials. This example, therefore, emphasises the need for a granular set of indicators to quantify and optimise CW from the early design stages, as included in PRODCIRO. Besides, the literature identified that time and cost constraints hinder the integration of CW into design and construction workflows [1]. So, with high information accuracy, accessibility, and granularity, the proposed workflow would allow the project team to make timely and integrated CW-informed decisions from early design stages. Furthermore, incorporating LCA stages achieved granularity in reporting the CW generation data across the life cycle stages. The systematic categorisation will facilitate decision-makers in identifying the impacts of each life cycle stage and implementing appropriate measures to reduce the impacts.

7.2.2 Standardised CW Data Ensure Interoperability and Automation

In addition to the breadth and depth of the information ('what type of CW data') the CWQ model holds, the following prominent challenge identified was 'how' the data should be standardised and structured to incorporate into the analysis. The existing CW platforms require high human intervention and significantly lack communication with the design and construction platforms (discussed in Section 4.4.5.2).

This, therefore, led to increased time and cost and reduced chances of comparing multi-design options before actual site work begins. Hence, utilising data standardisation, seamless data exchanges, and linked data attributes of SWT, the proposed BIM-integrated semantic framework is aimed to ensure interoperability and automation throughout the CW quantification and material optimisation process. The automated quantity estimation, life-cycle information management, parametric modelling, and API features of the BIM platform increased the information quality, quantity, accuracy, reliability, and dynamics of the developed framework. Besides, the logical inference and standardisation functionalities of SWT enhanced the integration of several CW data sources with BIM, thus enabling seamless information flows between the BIM and CWQ modelling environment. Thus, the standardised and interoperable CW data achieved by integrating BIM and SWT allow the AECO industry to utilise the full potential of ICTs in the CW domain and develop waste-efficient designs with less human effort and time. Besides, this data would act as a catalyst to overcome some of the management-related critical gaps identified from the stakeholder interviews (discussed in Section 4.4.5.4) and literature [2–5]: collaboration and communication and stakeholder engagement, involvement, and education across the supply chain.

7.2.3 Diverse CW Measurement Units Ensure Integrated Decision-Making

The other influencing theme discussed in the literature and interviews is the use of standardised units of waste measurement. The current CWQ models employ various measurement units, including weight, volume, generation rate estimation, and percentage analysis. However, the design team members stress that having a standardised unit aligning with other sustainability estimations could increase the chances of integrating CW into decision-making (discussed in Section 4.4.5.3). For instance, participant P10 emphasised:

> As a decision-maker in the design team, I am interested in having more granular data about what impacts the material waste would create and how critical it is to meet the sustainable targets in addition to knowing what amount of waste will be generated from a specific design.

Furthermore, DEFRA [6] also promotes impact-based waste analysis over volume- and weight-based estimations to align with net zero carbon targets and capital budgets. Although it is vital to introduce an impact-based analysis, the weight-based results are also equally crucial for the construction and WM team to provide adequate resources for waste storage, handling, treatment, disposal, and reporting. Thus, considering the multi-perspective dimensions of design, construction, and WM teams, the proposed framework integrated the impact-based assessment into the CWQ to ease design optimisation without compromising the benefits of weight-based estimations for the on-site team (presented in Section 6.4.3). This integrated system, therefore, eases the whole assessment process and maximises the chances of collaborative decision-making within the CW domain.

7.2.4 Uniclass Pr Codes Broaden the Material Optimisation Choices

Incorporating the Uniclass classification system into the CWQ workflow improved the optimisation process by providing expanded material choices to make informed decisions. Data mapping, the critical step within the optimisation process, is supported by Uniclass Pr codes in the developed system. Since the Uniclass codes follow a hierarchical approach (discussed in Section 6.3.3), object-based classification (fourth tier) is assigned to the instances of ontology classes to enhance the granularity of the material information. However, mapping BIM-based material information with product types in the ontology is established using the subject (third tier) to provide wider material choices in the early design stages. For instance, Section 6.4.3 presents that the clay bricks named 'GSS_ Brick Red Multi-stock' (Pr_25_57_06_15) in the external wall of the baseline design are replaced by *AggregateBricks_3* (Pr_20_93_52_02) in the optimised design. Although the baseline design states that it is a clay brick, it is more important to show the decision-maker multiple material choices that meet the functions of bricks (Pr_20_93_52, called 'Masonry walling units'). Therefore, using the subject-based classification, the materials that come under the Pr_20_93_52, including clay bricks, concrete bricks, recycled, and reclaimed bricks, are evaluated to optimise the most suitable one for the design. By not confining the choices only to a specific material type (in this case, it is clay brick), the decision-makers are provided with multiple options under this optimisation approach. Furthermore, the digitalised information about the alternative material choices and their technical and waste factors obtained from the framework ensure standardisation and consistency. As a result, the chances of automation and interoperability are increased, thus aiding in informed decision-making from the early design stages.

7.2.5 Semantic Relationships Unify the Difference Between Material and CW Characterisation Codes

Despite the fact that the coding system enhances accuracy in the CWQ results, there was a lack of analysis of 'how' to overcome the differences between the material and CW characterisation codes. For instance, the research conducted by [7, 8] linked the BCCA database and EWC to enable linear communication between the BIM material library and CW data. However, these integrations are database-specific and lack defined semantic relationships (discussed in Section 2.7.5). Besides, the coding approach used for material descriptions and waste classification does not align with each other. For instance, this research adopted Uniclass and EWC to enable seamless information flows between BIM and CW data models. In this regard, it was identified that EWC classifies waste materials based on their reactive nature. At the same time, Uniclass Pr codes classification in the AECO follows a granular approach based on their raw material composition, manufacturing process, and applications of the product type. Due to significant differences in the classification approaches between Pr and EWC codes, the complexity associated with tracking the material flows and making informed WM decisions increases. The inconsistencies with material and waste codes are also highlighted as one of the significant limitations impacting the AECO industry in achieving robust and timely CW data [9]. Thus, utilising the SWT functionalities, the developed CW data model provides a unified structure by establishing comprehensive semantic relationships between building materials and CW data (presented in Section 5.2). This unification aims to enhance integration and interaction across the different CW platforms, thereby reducing the inconsistencies in tracking the input and output material flows. Furthermore, the granular data extracted from the unified model allows stakeholders to make informed decisions and communicate effectively with internal and external stakeholders.

7.2.6 Proposed CWQ Framework – An Integrated CW Data Hub Supporting BIM Level 3 Adoption

IFC, one of the open-data schemas, has numerous benefits compared to proprietary data schemas. It extends accessible communication of project-related information back and forth between BIM and IFC-compatible software platforms, encouraging BIM Level 2 adoption (discussed in Section 3.2.5). Notwithstanding the benefits of IFC use, some barriers hinder its application in real-world scenarios. The IFC lacks formal semantics to map entities and the relationship between IFC classification and the data schema of the selected application. This results in the development of multiple fragmented data models containing the same information about a project but with different semantic definitions. Even though the IFC models enhance interoperability in the AECO domain, data standardisation and consistency are still significant shortcomings of the open-data models. Thus, the developed RDF model encourages a collaborative approach and the adoption of an integrated data repository by overcoming the limitations of existing open-data models. Furthermore, a machine-readable standard data structure with consistent taxonomies that map concepts between various information systems supports the BIM Level 3 adoption strategies. Therefore, accounting for the existing data limitations, the developed ontology would act as an integrated CW data hub aiding standardisation, integration, and collaboration across the life-cycle stages.

7.3 Summary

This chapter discussed the potential of the proposed CWQ workflow observed during the design, development, implementation, and validation of the framework. This includes the following.

- The accuracy and relevance of CW quantification and material optimisation can be achieved using a granular set of CW indicators.
- Maintaining a unified data structure and structured information flow enhances interoperability and automation across the quantification and optimisation workflow.
- The CWQ framework ensures that the decision-making is integrated and collaborative by using diverse CW measurement units.

- The material optimisation choices are broadened by categorising the building products using the Uniclass classification system.
- Semantic relations established in the ontology unify the differences between material and CW classification codes.
- The ontology demonstrated using the proposed BIM-semantic framework would act as an integrated CW data hub supporting BIM Level 3 adoption.

References

1. S. Sivashanmugam, R. Sergio, R. Farzad and D. Nashwan, Maximising the construction waste reduction potential – Barriers and catalysts. *Proceedings of the Institution of Civil Engineers – Civil Engineering*, 2022. 176(6): pp. 1–9. https://doi.org/10.1680/jcien.22.00163
2. E.Y. Frempong-Jnr, S.O. Ametepey and J.E. Cobbina, Impact of stakeholder management on efficient construction waste management. *Smart and Sustainable Built Environment*, 2023. 12(3): pp. 607–634. https://doi.org/10.1108/SASBE-08-2021-0147
3. K. Krueger, A. Stoker and G. Gaustad, "Alternative" materials in the green building and construction sector: Examples, barriers, and environmental analysis. *Smart and Sustainable Built Environment*, 2019. 8(4): pp. 270–291. https://doi.org.ezproxy.tees.ac.uk/10.1108/SASBE-09-2018-0045
4. M.B. Purushothaman and J. Seadon, System-wide construction waste and their connectivity to construction phases, impacting 5M factors and effects: A systematic review. *Smart and Sustainable Built Environment*, 2024. 13(2): pp. 354–369. https://doi.org/10.1108/SASBE-12-2022-0269
5. X. Zhao, Stakeholder-associated factors influencing construction and demolition waste management: A systematic review. *Buildings*, 2021. 11(4): p. 149. https://doi.org/10.3390/buildings11040149
6. Defra, *Our waste, our resources: A strategy for England*, 2018. https://assets.publishing.service.gov.uk/government/uploads/system/uploads/attachment_data/file/765914/resources-waste-strategy-dec-2018.pdf. Last access: 02 April 2023.
7. R. Quiñones, C. Llatas, M.V. Montes and I. Cortés, A multiplatform BIM-integrated construction waste quantification model during design phase. The case of the structural system in a Spanish building. *Recycling*, 2021. 6(3): p. 62. https://doi.org/10.3390/recycling6030062
8. R. Quiñones, C. Llatas, M.V. Montes and I. Cortés, Quantification of construction waste in early design stages using BIM-based tool. *Recycling*, 2022. 7(5): p. 63. https://doi.org/10.3390/recycling7050063
9. Construction Leadership Council, *Zero avoidable waste in construction*, 2020. https://www.constructionleadershipcouncil.co.uk/wp-content/uploads/2016/05/ZAW-Report-Final-Draft-25-February–2020.pdf. Last access: 15 May 2023.

8 Conclusions and Future Directions

8.1 Introduction

Globally, half of the raw materials and one-third of solid waste generation result from built environment activities. In recent years, the disposal of CW in landfills has been reduced, but the amount of waste generated in the first instance is still high. Several technological advancements, process changes, and policy updates have been introduced to manage the CW generation levels in the built environment. However, the literature emphasises that accurate quantification of CW in the early design stages is crucial to minimise their generation. Besides, it is also acknowledged as a prerequisite for effective management of CW across the life cycle stages. Over the last decades, various quantification models have been developed to measure, manage, and report construction waste generation. A comprehensive SLR of research studies identified that the CWQ models primarily focused on enhancing the CW quantification and management process with a limited focus on data and information standards. Advanced ICTs like BIM, AI, big data, GIS, and robotic designs were used by the existing CWQ models to enhance the speed and accuracy of quantification. However, the lack of standardised and unified CW data across these models limited their wide-scale application. Furthermore, it was identified that this data gap limits the built environment from utilising the full potential of advanced ICTs to quantify and optimise CW from early design stages.

After identifying the knowledge gap, the research sought to propose a semantic framework underpinned by BIM, SWT, and LCA integration. The critical realistic philosophical paradigm supported the design of the research methodology. A six-step flexible DSR approach is implemented to develop, apply, and evaluate the CWQ framework. The application of the framework was demonstrated using a semantic CW data model and a BIM-integrated digital tool. The framework aims to ensure that the four main dimensions of the CWQ process, including 'interoperable and standardised data', 'unified data structure', 'automated quantification and optimisation', and 'consistent classification and reporting', are achieved using the proposed framework. Besides, this framework is expected to support the built environment in shifting from 'reactive CWM' to 'proactive and preventive CWM'.

DOI: 10.1201/9781003610762-8

8.2 Interpretations and Conclusions

The interpretations and conclusions drawn from the design, development, and validation of the CWQ framework are discussed in the following sections.

8.2.1 Current Scenario of CW Generation and Management

An increase in built environment activities increases raw material consumption, energy usage, and waste generation globally. CW generation is a result of various activities across the project life cycle. The factors associated with design, procurement, construction, management, site operation, and external factors significantly influence waste generation. However, the factors associated with the design phase contribute to one-third of the total CW generated. Besides, the design stages have higher chances of preventing waste from being generated. The excessive generation of CW will have a detrimental effect on the environmental, economic, and social sustainability pillars. The use of advanced technologies, changes in the supply chain and waste handling process, expanding the use of secondary products, and increased regulatory policies support waste reduction and management across the project life cycle stages. Besides, the reuse and recycling process reduced the quantity of waste disposed of in landfills. However, the current CWM system is fragmented and lacks a systematic approach. So, a systematic CWM approach, with clear waste targets and enforceable legal standards, is crucial to enhance CW reduction and reduce burdens CW puts on achieving a sustainable built environment.

8.2.2 Applications of Existing ICT-Integrated CWQ Models

Quantifying the amount, time, source, and impact of CW generation is a prerequisite for developing comprehensive WM plans. Several models and frameworks have been proposed to measure waste generated from construction, refurbishment, and demolition activities. The applications of advanced ICTs like BIM, AI, GIS, big data, AR, VR, and blockchain to quantify and classify CW is also increasing. Integrating these technologies increases the accuracy of CWQ, reduces the level of human intervention associated with the CWQ process, and mitigates errors arising from manual data handling. However, the data and information gaps limit the adaptability and scalability of existing CWQ models. Some limitations are as follows. First, the CW data recorded across most existing CWQ models are generic and lack explicit semantic relations with waste-causing design and construction factors. So, the non-granularity and non-specificness of CW data hinder its use across other applications. Second, the focus of existing models is limited to a specific project life cycle stage, so there is a high chance for the impacts to be shifted to other stages. Third, most existing CWQ models are unidirectional and lack dynamics with the BIM-based design and construction platforms. As a result of these limitations, the full potential of CWQs cannot be realised. So, in addition to automating the process, the consistency and granularity of CW data are vital to achieving the most from the ICTs. Besides, the quantity of waste generation measured in weight and volume is used to make CWM decisions. However, integrating the impacts of generated CW is vital to making sustainable decisions and meeting net-zero targets. Hence, these concluding factors form the basis for developing a framework that promises standardisation, interoperability, consistency, and dynamics across the CW quantification and optimisation workflows.

8.2.3 Attributes of BIM, SWT, and LCA Supporting CWQ

The functionalities of BIM, SWT, and LCA show great potential for streamlining and automating the CWQ workflow. A comprehensive literature review approach was used to meet the objective. The applications of SWT and BIM to develop an integrated approach that would overcome the existing limitations were studied. Besides, the adoption of a systematic approach guided by LCA and their applications in enhancing the CWQ workflow was reviewed. The attributes of BIM, SWT, and LCA supporting CWQ include the following:

- BIM promise a granular and dynamic workflow between the CW and BIM platforms. The main attributes supporting the process are life-cycle information repository, automated quantity extraction, parametric modelling, visualisation, interoperability, integration, and APIs. In addition, BIM LOMDs and maturity levels significantly influence the quantity and quality of CW information.
- The functionalities of SWT, including standard data representation and exchange, logical semantic inference, and seamless integration of semantic web data with BIM, have potential applications in the CWQ process. The RDF models ensure data standardisation, unification, semantic relation(s), consistent waste classification, and structured and seamless information flows.
- The attributes and functionalities of LCA, including its focus on the whole life cycle, use of comprehensive impact indicators, and data-driven decision-making, would be a potential solution to propose a holistic CWQ framework. It is also concluded that the LCA approach enhances the transparency and granularity of CW results.

8.2.4 Actual CWM Scenario in the Industry and Future Expectations – AECO Stakeholder's Perspectives

Since built environment stakeholders are end users of the CWQ framework, understanding their perspectives on existing scenarios and future expectations was crucial. Besides, integrating their perspectives was vital to enhance the real-life applications of the CWQ framework. The stakeholders involved in the decision-making of CW across the project life-cycle were called for interviews. The stakeholders emphasised the following main aspects:

- BIM supports CW reduction factors like clash detection, design detailing, and communication; however, the lack of waste data at the design stage limits designers and architects from integrating CW quantification and material optimisation processes into the actual project workflows.

- In addition to 'what type of CW information is available', 'how the CW information is integrated into the design process' is crucial for developing waste-efficient designs.
- Integrating the GWP values of CW is essential for making informed decisions in the design phases.
- Collaboration and communication among the supply chain members significantly influence CWM across the life-cycle stages. However, enhancing the data and technology adopted is at the forefront of achieving this collaboration.
- Among the existing CWM options, reactive measures like reuse and recycling are more prevalent over proactive and preventive waste reduction measures in the industry.
- In addition to WGR, a granular set of technical, environmental, and economic factors of materials are critical for making CW-informed decisions in the design stages.

8.2.5 *Features and Functions of BIM-Based Semantic CWQ Framework*

The conclusions drawn from the literature review and semi-structured interviews led to the development of a holistic framework that supports the built environment to quantify CW and inform optimal material choices. The proposed BIM-integrated semantic framework facilitates CWQ from the early design stages. The main conclusions obtained during the design and development of the framework include the following:

- The systematic and dynamic approach should be a primary functionality within the CWQ framework.
- Three main modules are vital to streamlining the CWQ workflow. It includes the data collection/extraction module, the data mapping and standardisation module, and the BIM-based CWQ and decision-making module.
- Enabling integrated digital systems, ensuring standardisation and consistency in information flows, and ensuring that the overall CWQ workflow is interoperable and automated are vital to meet the digital demands of the built environment.
- Recording CW data is vital to enhance the efficiency of CWQ and CWM.
- The quantity and quality of inputs decide the accuracy and granularity of CWQ results.

8.2.6 *Enhanced Interoperability and Automation Using PRODCIRO and BWT*

The application of the proposed framework is demonstrated via an ontology (PRODCIRO) and a BIM-integrated digital tool (BWT). The linked data functionalities of SWT and BIM APIs were used to demonstrate the application. The PRODCIRO is a metadata schema that provides explicit specifications of CW entities and relationships among them to support a comprehensive CWQ process. Besides, the BWT ensures seamless information flows between the BIM and PRODCIRO, thus ensuring the overall CWQ workflow is interoperable and automated. Besides, the UI is integrated to support designers and architects in accessing the CW data and tracking design changes in real time. Demonstrating the framework's application using PRODCIRO and BWT concluded the following:

- As a result of compromised data quality and transparency, existing CW databases encounter compatibility and reliability issues. So, there is a need to integrate various knowledge sources to develop a detailed semantic data model applicable across the CW domain. Besides, depending on multiple knowledge sources is desirable over a single source to reduce data gaps and enhance wide-scale applications. So, various knowledge sources, including literature reports, interview data, on-site waste records, CW databases, life-cycle impact data inventories, and standard classification systems and guidelines, are used to model an ontology.
- The CW variables selected for decision-making might vary with building type, geographical location, material sources, and client requirements. The extent to which these requirements could differ might be dynamic, so a granular model should accommodate all these necessary factors during design, development, and implementation.
- Adopting the functionalities of semantic RDF/OWL and SPARQL, highly granular open data models can be established to structure information flows, ease data exchange, and enhance interoperability and automation with the CW domain.
- Despite the data standardisation and unification achieved using the ontology, integrating the semantic waste-data model with BIM designs ensures automation, application, timeliness, and easy decision-making.
- Although SPARQLs enhance the ontology services, writing long queries is a time-consuming and human-intervened task. This limitation could be mitigated by incorporating the advanced RDF libraries into the Dynamo Revit, thus preventing users from managing multiple data models.
- Streamlining the overall CW quantification and material process within BIM could be achievable via a user-interactive digital tool. This process is generally considered complex in the literature and interviews because of multiple interconnected factors. So, it is vital to ensure accessibility and ease of working with the developed tool to allow the design team to make timely waste data-driven decisions, increase transparency in waste reporting, and enhance communication and collaboration within the supply chain.
- Uniclass classification systems broaden the optimisation choices. It also allows designers and architects to understand the performance of other materials, compare various materials, and explore different material sources, given that the materials meet the desired application.
- The technical functionalities of the material should always be prioritised during the optimisation and decision-making.

8.2.7 Accuracy and Adaptability of the CWQ Framework

The accuracy of CWQ results and adaptability of the proposed framework within the existing project workflows are tested using a test case building. The interpretations made at this stage are as follows:

- Semantic data models ease data exchange, structure the waste information flows, and provide a granular set of CW information to ease decision-making from the early design stages.
- Making conclusions about CW should account for multiple factors across life-cycle stages. For instance, a material is considered less waste-generating in the construction/installation phases, but its impact on manufacturing and EoL stages might be higher than that of other materials. Therefore, a detailed analysis of each CW indicator should be provided to the user so that they can make informed decisions.
- The most influencing optimisations are changing the material type, shifting the material source (raw/virgin to secondary sources), and selecting an alternative manufacturer without changing the material type and source. Since each of the discussed optimisation methods has varying levels of impact, it is up to the user to apply the most appropriate material based on the design requirements.
- At present, the material choices informed as optimal are based on the CW generation and their associated GWP values. The choices may change if other impact indicators are accounted for in the framework.

8.3 Research Implications

This research has significant implications for streamlining the CWQ workflow from early design stages. Each of the theoretical and practical implications is discussed in detail.

8.3.1 Theoretical Implications

By integrating BIM, CW, and life-cycle impact data repositories, the proposed framework makes the quantification and optimisation process more streamlined and effective. Although the importance of CW is realised globally, a paucity of clear indicators hinders its management. So, the research identified 22 different types of generic, technical, environmental, and economic factors that should act as substantial indicators for proposing a sustainable WM solution and developing waste-efficient designs. Thus, using this granular set of CW information with precise semantics, the proposed framework would act as a standardised CW data catalogue. Besides, it enables a shift from presenting generic non-classified waste generation levels to a more precise and detailed material-based classification, facilitating the whole life-cycle waste impact analysis from the early design stages.

Furthermore, integrating the semantic model into BIM enables bidirectional communication between BIM models and ontology datasets. This also proved the importance of having standardised units and shared data structures between BIM and other databases to achieve a fully integrated and interoperable workflow. The research outcomes, therefore, support some of the critical pillars of digital innovation: structured information flows, eliminating data silos, unified data structures, and interoperability. Also, the development and validation of the proposed framework proved that the data is vital to maximising the efficiency of ICTs, achieving a collaborative system, easing communication, and making timely decisions.

8.3.2 Practical Implications

Designers and Architects are the primary beneficiaries of the research outputs. With increased focus on design stages to reduce CW generation, designers and architects are encouraged to develop waste-efficient designs. Although various process changes are adopted to enhance CWM, interview results showed that reliable waste data is vital to enhance decision-making. Thus, the semantic framework proposed in the research provides accurate and consistent CW data that is interoperable with the BIM platform to make informed decisions. Besides, the ontology and the BIM-based tool ensure that the data from the design and CW domains are linked seamlessly to make decisions. This, therefore, facilitates designers and architects to quantify waste, optimise waste-efficient materials, benchmark waste levels, and compare multiple design options in a virtual environment in no time.

Sustainability Professionals – Playing an active role in the design and construction phases, sustainability professionals aim to set targets for each building to minimise material waste and maximise circularity. So, one of the critical barriers confronted in the current system is the lack of quality data to set standardised waste targets. So, the waste and its impact results from the research would support them in setting clear targets and bridging linear communication between the design and construction teams.

Policymakers – Similar to sustainable professionals, policymakers are actively involved in setting building and national-level targets to reduce the impacts of CW. Despite the existing CW targets, reports suggest a need for impact-based waste targets to reduce carbon and cost. These targets further support the industry in sourcing sustainable materials, minimising waste, and selecting appropriate waste treatment routes to enhance circularity levels and reduce whole-life carbon emissions.

Besides, this integrated model with precise semantics allows (i) *researchers* to utilise the information to develop machine-learning models, model robotics, and facilitate the WM process at a broader scale, and (ii) support *product manufacturers* to measure the CW impacts of their products and compare the results with their competitors.

8.4 Knowledge Contributions of the Book

8.4.1 Contributions to CW Research Agenda

i) The workflow proposed in the research is the first attempt to integrate BIM, CW, and life-cycle impact data semantically. Previous works focused on the CW quantification and classification 'process' with limited focus on the 'data'.

This research focused on enhancing the 'CW data structure and standards' that support the entire 'process'.

ii) A bidirectionally connected framework is proposed to enable seamless information flows between the BIM and semantic CW data models.
iii) A semantic data model is developed with a granular set of technical and environmental information to enhance the integration of the CW-AECO domain.
iv) The UI ensures that data extraction, material mapping, quantification, optimisation, and visualisation of waste results are automated and integral to BIM.

8.4.2 Contributions to United Nations Sustainable Development Goals

Integrating the dimensions of environmental sustainability and innovational approaches within the AECO-CW domain, the research contributes to some of the United Nations Sustainable Development Goals. Around half of the global solid waste is generated from built environment activities. Excessive waste generation leads to increased consumption of raw materials and energy usage. So, there is a pressing need to prevent the waste generation, reuse the waste products and minimise the raw material consumption. The framework proposed in the book is a holistic solution that facilitates the built environment to predict the quantity of waste a building is expected to generate across its life cycle. Besides, the application of the framework in the design phase will enhance waste prevention and inform optimal material choices by accounting for CW generation and impacts. This comprehensive optimisation approach ensures that the built environment will consume less raw materials, increase the use of recycled and renewable materials, select durable materials, and enhance waste reuse and recycling. This, overall, contributes to UN SDG 12 – Resource Consumption and Production, where minimising the material and energy footprint levels is promoted. Furthermore, the framework accounts for the carbon emissions of CW while optimising material choices. As a result, decision-making supports minimising the global carbon footprint, as Goal 13 – Climate Action emphasises.

Besides, pollution and landfilling are some of the significant environmental challenges that result from CW generation. The increased pollution affects the health and well-being of individuals. The landfilling of waste also impacts the entire ecosystem. These challenges can be managed by optimising materials and activities that generate less waste, as supported by the framework. Thus, the design and implementation of the framework support the built environment to contribute to the objectives of UN SDG 11 – Sustainable Cities and Communities and UN SDG 3 – Good Health and Well-being. The technological and sustainable dimensions of the framework also foster sustainable innovation within the built environment, thereby contributing to the goals defined in UN SDG 9 – industry, innovation, and infrastructure.

8.5 Current Limitations and Future Research Agenda

The proposed framework focused on estimating the GWP impacts of waste materials across the life-cycle stages. Although the research adopted material-specific data to estimate the GWPs, the transportation and waste treatment scenarios are extracted directly from EPDs, excluding the differences between these assumed values and actual scenarios on-site. Therefore, future work in this direction should explore and integrate real-life site scenarios in decision-making. Adopting image-based models to capture real-time data into the ontology could enhance the integration of actual scenarios and increase the accuracy and relevance of results. Besides, the impact of waste is across the triple bottom line of sustainability: environmental, economic, and social. Hence, future research should demonstrate the integration of three sustainable pillars while making CW-related decisions. A multi-criteria decision-making model integrating environmental, economic, and social impacts would be a promising solution. Although the proposed framework holds wide-scale applications, its application, which is tested using an ontology and a BIM tool, depends on UK-based scenarios. So, the influencing waste indicators and impact-based results might differ from other geographical locations.

Another potential research area could be integrating AI elements with the developed BIM-based semantic system. Since the existing semantic waste model provides structured and consistent waste data, incorporating the machine-learning models could support the researchers to reduce inconsistencies and gain valuable insights, thereby utilising the full potential of AI for CW quantification, classification, optimisation, and even support robotic on-site waste picking and sorting. Data is one of the critical pillars of the AI process, so the developed ontology with standard semantics could support research to obtain quality CW data for modelling. Besides, the developed tool integrates with 3D design and material information without accounting for the factors of time and space on-site that may influence waste generation levels. Therefore, future research could explore the possibility of integrating these factors into the AI-based CWQ workflows.

Index

Note: Page numbers in *italics* indicate a figure and page numbers in **bold** indicate a table on the corresponding page.

For Product Safety Concerns and Information please contact our EU
representative GPSR@taylorandfrancis.com
Taylor & Francis Verlag GmbH, Kaufingerstraße 24, 80331 München, Germany

www.ingramcontent.com/pod-product-compliance
Ingram Content Group UK Ltd.
Pitfield, Milton Keynes, MK11 3LW, UK
UKHW022036061126
477531UK00008B/360
* 9 7 8 1 0 4 1 0 0 6 0 9 1 *